Knowledge Organization and Management
in the Domain of Environment and Earth Observation
(KOMEEO)

T0134348

Advances in Knowledge Organization, Vol. 18 (2022)

Knowledge Organization and Management in the Domain of Environment and Earth Observation (KOMEEO)

Proceedings
of the
International KOMEEO Conference
2021
Rende, Italy

Organized by
the Laboratory of Documentation,
Department of Culture, Education and Society –
University of Calabria
and by
the Italian Chapter of the International Society
of Knowledge Organization (ISKO)

Edited by

Antonietta Folino
Roberto Guarasci

ERGON VERLAG

Editorial Support:

Giovanna Aracri, Institute of Informatics and Telematics –
National Research Council (IIT-CNR)

Armando Bartucci, University of Macerata

Susie Caruso, University of Calabria

Martin Critelli, University of Calabria

Maria Teresa Guaglianone, Institute of Informatics and Telematics –
National Research Council (IIT-CNR)

Claudia Lanza, University of Calabria

Erika Pasceri, University of Calabria

The publication has been funded within the ERA-PLANET Program –
The European network for observing our changing planet –
Call: H2020-SC5-2015-one-stage; Topic: SC5-15-2015;
Type of action: ERA-NET-Cofund; Grant Agreement n. 689443.

Bibliographic information published by the Deutsche Nationalbibliothek:
The Deutsche Nationalbibliothek lists this publication in the Deutsche
Nationalbibliografie; detailed bibliographic data are available in the
Internet at http://dnb.d-nb.de.

Published by Ergon – ein Verlag in der Nomos Verlagsgesellschaft, Baden-Baden 2022
Overall responsibility for manufacturing (printing and production) lies with
Nomos Verlagsgesellschaft mbH & Co. KG.
Cover Design: Jan von Hugo

www.ergon-verlag.de

ISBN 978-3-95650-874-5 (Print)
ISBN 978-3-95650-875-2 (ePDF)
ISSN 0938-5495

Table of Contents

EraPlanet and KOMEEO

Introduction

Roberto Guarasci
University of Calabria, Italy

In 1995, Peter Ingwersen (Ingwersen 1996) developed an epistemological map of the interconnections between the various information sciences, placing communication, cognition and systems sciences at the highest level, which he defined as interdisciplinary, while he placed documentation science and librarianship at an applicative level, thus re-proposing the unresolved problem of the relationship between science and technique. This relationship is particularly significant in documentation science in which the applicative content of some theoretical assumptions is an important part of the cognitive process. If, however, each technique is associated with a science that determines its theoretical and conceptual foundations, which the former merely makes operational, which science could archiving, documentation and librarianship rely on?

> *"Ainsi la 'technique documentaire' serait dépendante d'une science, celle que Yves F. Le Coadic appelle la 'documentation' ou la 'science du document', celle que quinze ans plus tôt était pour Robert Escarpit la 'documentologie', celle qui pour beaucoup est aujourd'hui identifiée sous l'expression 'sciences de l'information'"* (Fondin 2002, 122).

Techniques are a set of practical knowledge, of know-how, typical of the trades or arts in which non-formal learning plays a major role and allows one to create or transform objects. Each technique, more or less knowingly, uses methodological and theoretical assumptions, typical of one or more sciences, in a practical manner. These are then translated into operational reality thanks to one's ability, thus giving the finished product a certain originality, which derives from individual characteristics, albeit in the presence of standardized procedures and paths. Artistic craftsmanship is, in this case, the most emblematic example. A goldsmith uses techniques whose theoretical assumptions lie in the physical, chemical and mechanical sciences, not necessarily having full theoretical knowledge but mastering the operational derivation in a customary fashion. Just as a Stradivari violin is not the simple sum of its components, so does each individual skill give added value and originality to this type of product.

The relationship between technique and science is not always univocal but often multiple: one to many. In the documentation sciences, it is undoubtedly true that there are significant technical components and that, consequently, a documentation technique can be identified which includes all the necessary skills to draw up a catalographic record, extract the terms to be used in a taxonomic structure, or reconstruct the original order of an archival fonds. But which science or sciences does it refer to?

If we look at it from another point of view, for a science to be considered such, it is necessary that it identifies one or more objects of study, theories and laws and its own domain lexicon capable of univocally defining the objects in a community of use. The object we refer to in documentation science is the document in its various meanings and constituent elements. If we use Suzanne Briet's definition:

"Une étoile est-elle un document ? Un galet roulé par un torrent est-il un document ? Un animal vivant est-il un document ? Non. Mais sont des documents le photographies et les catalogues d'étoiles, les pierres d'un musée de minéralogie, les animaux catalogués et exposés dans un zoo" (Briet 1951, 7), reference sciences are far reaching, but even if we use a much more restrictive and defined one, which identifies the document as "information created, received, and maintained as evidence and information by an organization or person, in pursuance of legal obligations or in the transaction of business" (ISO 2001), certainly, it cannot be claimed that an object of study is unique to a single science.

Moreover, when in 1951 the World Congress of Documentation was held for the first time in Rome, it was noted that:

> *"Ed è una cerchia ben ampia e variata di persone quella che il documentalista è chiamato ad aiutare con la sua tecnica: l'industriale, lo scienziato, il tecnico, il ricercatore di laboratorio, il medico, lo studente, l'ingegnere, il commerciante; in alcuni paesi, come in Francia, il documentaliste è ormai un professionista, in Inghilterra l'information officer è un tecnico, ha già dei corsi da seguire, delle scuole da frequentare, dei diplomi da esibire"* (Pinto 1952, 4).

When, further on, a census of the documentation bodies in Italy was carried out on the basis of a survey conducted by the Centro Nazionale di Documentazione Tecnico Scientifica of the CNR (National Center for Technical and Scientific Documentation of the CNR)[1] almost all of the individuals belonged to large industrial groups or entrepreneurial associations, from Dalmine to the Istituto Siderurgico FInsider to Assofond, the Associazione Nazionale delle Fonderie. In all cases the activity was that of collecting and

---

1   On the History of the National Center for Scientific Documentation of the CNR see (Guarasci 2011).

making bibliographic and patent material on the specific field of competence available by filing individual documentary units and bibliographic resources according to the method widely tested by the International Bibliographic Institute and propagated in Italy by the abovementioned Centre of the National Research Council.

In the meanwhile at the International Bibliographic Institute the filing was directed towards the creation of the Universal Bibliographic Repertory, the operational realization of a broad theoretical reflection on the universalism of knowledge and on the role of books and documents in Italy, probably also because of the deep aversion to the Universal Decimal Classification *"scarsamente conosciuta, ancor più scarsamente adottata e ciò che più nuoce quasi mai applicata integralmente e fedelmente"* (Ascarelli 1952, 45)[2], what was almost exclusively emphasized was the technical and instrumental aspect that, with the arrival of mass computer science, represented its weakest point and determined its rapid decline, placing the other documentation sciences on contents and methods that often – in the past few decades – have refused any type of contamination in the attempt to defend a delimited specificity.

In some European countries such as France, documental techniques had documentation science as a reference point, the latter has naturally evolved and has borrowed and fused theoretical assumptions coming from information science, along with its cognitive baggage, giving rise to a new Documentation Science. It would be more correct to call it – as some do – documental engineering (Guyot 2012). In other European countries, the emphasis on technique and the strict boundaries between disciplinary fields have often led to the creation of links between those techniques and the sciences appertaining to the application domains, so that the biomedical documentalist is drawn to the sphere of pharmaceutical sciences, the patent documentalist to hard sciences and so on.

If the ultimate goal of *"offrir sur tout ordre de fait et de connaissance des informations documentées"* (Otlet 2015, 6) – stated by Paul Otlet in the Traité – seems to have been completely absorbed by information science, the subsequent specification of the *"parties de la documentation"* which includes – in a synergic and complementary vision – libraries, archives and museums is quite topical for the contamination and the synergies it presupposes about the object of study:

> *"Le document est l'objet d'un Cycle d'opérations réalisant la plus complète division du travail et l'utilisation la plus disperse de ses résultats [...] Il devient l'objet d'un travail complémentaire tendant à le juger et à l'apprécier, à en incorporer les données particulières aux données déjà existantes"* (Otlet 2015, 7).

---

2   On this subject see (Fumagalli 1896).

When, in 2014, the idea of a project for the construction of a European network for the observation of our changing planet slowly took shape (ERA-PLANET, http://www.era-planet.eu/), one of the first critical elements that emerged was precisely the need for a deep synergy among skills to try to overcome the limited availability of multilingual terms to define the Essential Variables within the Societal Benefit Areas with the consequent problematic match between the information requests of users, experts and producers of the data. It was deemed necessary to start from existing terminology and classification structures, make them homogeneous, integrate them and create a comprehensive and interoperable knowledge base.

> "As a multitude of heterogeneous data will be made available through the GEOEssential Knowledge Base infrastructure, it is essential to ensure high standards of discoverability, accessibility, and interoperability. The design of this infrastructure involves the alignment and integration of a set of semantic resources defining the specific domain. The latter is important in order to ensure harmonised access to the vast volume of data produced, turning it into usable information and knowledge, and to guarantee semantic interoperability within the infrastructure. This involves the mapping of existing aligned thematic vocabularies (i.e. glossaries, taxonomies, thesauri and ontologies), along with the integration of further domain-specific terminology obtained through a corpus-based approach. The integration of the abovementioned vocabularies in a knowledge base infrastructure will therefore improve the ability of end-users to explore and exploit EO data. Some of the vocabularies employed are the following: GEMET Thesaurus, INSPIRE Feature Concept Dictionary and Glossary, EARTh Thesaurus" (Folino, Caruso, and Aracri 2018, 2).

After six years of work, the goal has been achieved and this volume brings together the experiences of those who have contributed – in various ways – to making it possible, either through direct participation in the project activities or through the overall theoretical reflection that constitutes the fundamental and inevitable framework of that realization.

It is a reflection on the opportunities and possibilities offered by the operational contamination between documentation science, information science and domain knowledge, united by the attempt to build a new, symbolic, intermediate layer between terminology, documentation science and domain knowledge, capable of representing a cognitive bridge between the techniques and sciences involved. It is not an abstract contamination but a real sharing of choices, methods and techniques. In James Gleick's words, we have tried to stop being "colour blind". When Europeans in the 19th century tried to decode the informative value of the sound of drums, they tried to transliterate those vibrations into the Latin alphabet, not realizing that the secret was tone because, "drummers could not rely on an intermediate code because African languages did not have an alphabet. The drums metamorphosed the spoken language. [...] transliterating the words they heard into the Latin alphabet. They completely neglected tone. In fact, they were colour-blind" (Gleick 2012, 29).

References

Ascarelli, Fernanda. 1952. "L'Applicazione della CDU in Italia." In *La Documentazione in Italia, Atti del congresso mondiale di Documentazione*, Roma, Consiglio Nazionale delle Ricerche.

Briet, Suzanne. 1951. *Qu'est-ce que la Documentation ?*. Paris : Editions documentaires, industrielles et techniques.

Folino, Antonietta, Assunta Caruso, and Giovanna Aracri. 2018. *Deliverable 1.4, Semantic Services*. GEOEssential EraPlanet.

Fondin, Hubert. 2002. "La Science de l'Information et la documentation, ou les relations entre science et technique." *Documentaliste – Sciences de l'information* 39, no. 3: 122-9.

Fumagalli, Giuseppe. 1896. "La Conferenza internazionale bibliografica di Bruxelles e il repertorio bibliografico universale." *Rivista delle Biblioteche e degli Archivi*, anno VI, n. 9-10: 129-50.

Gleick, James. 2012. *L'Informazione*. Milano: Feltrinelli.

Guarasci, Roberto. 2011. "La Memoria della Scienza: L'Archivio Tecnico Italiano e il Centro Nazionale di Documentazione Scientifica." In *Archivi privati. Studi in onore di Giorgetta Bonfiglio-Dosio*, a cura di Roberto Guarasci and Erika Pasceri. 195-218. Roma: CNR – SeGID,

Guyot, Brigitte. 2012. *Introduction à l'ingénierie documentaire et aux sciences de l'information*. Paris : CNAM.

Ingwersen, Peter. 1996. "Cognitive perspectives of information retrieval interaction: elements of a cognitive IR theory." *Journal of Documentation* 52, no. 1: 3-50.

ISO 15489:2001, Information and documentation – Records management – Part 1: General, 3.15.

Otlet, Paul. 2015. *Traité de Documentation*, Mons : Les Impressions Nouvelles.

Pinto, Olga. 1952. "Introduzione a La Documentazione in Italia." In *Atti del congresso mondiale di Documentazione*, Roma, Consiglio Nazionale delle Ricerche.

Crowdsourcing classification and causality to power a search-and-innovation engine

Richard Absalom
Independent Researcher, Luxembourg

Abstract

Durham Zoo (DZ) is a project to create a search-and-innovation engine for science and technology. The engine has been designed to capture the knowledge of experts from different areas of expertise via the classification of the literature. The architecture combines the higher-level cognition of humans and their powers of language, abstraction, of inference and analogy, with the storage and processing power of computers. The system is adapted for searching both what already exists, and novel solutions to problems. To be built and operated by the community, the goal is to democratize innovation whilst funding societal causes such as climate-change mitigation or the search for new antibiotics. The original design, first published in Absalom and Absalom (2012), relied upon fuzzy and faceted classification. The fuzziness related to the similarity of concepts in each of the facets. A search query would be matched with the literature in multiple facets to retrieve holistically similar literature, or to suggest a solution to a problem from elsewhere in technology or the natural world. The facets used to describe a concept in science and technology included a problem and a solution. A recent reappraisal of the project design recognised the potential of causality for modelling and matching problems. This paper proposes a design compatible with the crowdsourced classification.

1.0 The original motivation for the project

The initial challenge was how to crowdsource the classification of the literature in support of patent searching. Patent offices still use classification by experts as a cornerstone to searching the prior art. For whilst information retrieval and artificial intelligence are making great strides, the human brain is still class leading at identifying and understanding concepts. It is still best able to work through imperfect language, abstraction, jargon, and terminology to distill the essence of a disclosure. The essence of analogous concepts is encoded with a same classification code. In the patent world there is the added complexity of the legal nature of a patent: things are often described in broad terms so as not to restrict the scope of protection. As an example, a magnetic disk drive may be described as a "storage device". This can complicate search using keywords.

Rapidly increasing numbers of patent applications and a massive increase in the scientific literature resulted in a scalability problem. Could automatic classification produce the goods? Could not patent applicants and authors better classify their own disclosures?

Much patent office classification is based to a greater or lesser degree on the International Patent Classification (IPC). The IPC is a fantastic resource, the result of the considerations of experts over many years. Unfortunately, the complexity, the classification rules and the esoteric patent-speak constitute a barrier to entry as regards a crowdsourcing effort. Could we not design something simple and intuitive?

The IPC's origins as a paper classification scheme with a hierarchical tree structure restrict its ability to evolve with technology. Digital convergence saw an increase in 'sameness' between computing, on the G root class, and telecommunications and television on the H root class. This was reflected by an increasing overlap between the two classes and much dual classification. Dual classification is not a problem *per se*, however creeping uncertainty and ambiguity in classes has consequences for precision and recall. The multidisciplinarity of nanotechnology, combining all manner of physical sciences, life sciences and engineering from across all the A to H root classes complicated matters further. Was the tree structure not the root of the problem?

The paper-classification origins have resulted in a limited use of faceting, perhaps with the exception of the Japanese Patent Office's electronic implementation of the IPC. The use of "on-the-shelf-or-not" Boolean classification fails to represent the degree of sameness of different concepts. This is better done with fuzzy classification. Fuzzy mathematics can return a ranked list of hits to a search query. What of a fuzzy and faceted classification scheme?

How to manage a complex classification scheme? The IPC required experts to assemble and discuss both the "what is what" and the 'what goes where" of new technologies. This requirement for centralised management is incompatible with a distributed crowd of individuals working independently.

2.0 The basic design for single concepts

The design process proceeded in ignorance of knowledge organisation theory and terminology. The terminology used here is neither the original terminology used, nor that adhering to a standard. However, the basic design has much in common with a faceted thesaurus and where possible consistency with the ANSI/NISO standard has been sought.

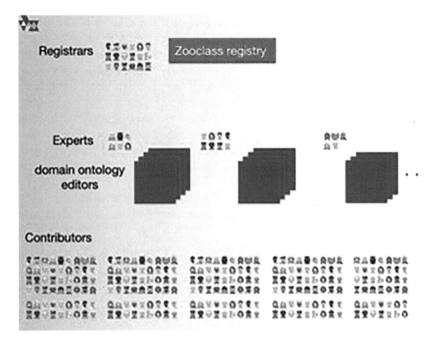

Figure 1. Three-tiered structure

The design has three actors, Registrars who manage a controlled vocabulary, Experts who use the controlled vocabulary to build domain ontologies, and a crowd of Contributors who are invited to classify and search the literature, see Figure 1.

It was decided that defining "what is what" with a controlled vocabulary was essential to the design: experts from different areas of expertise commonly interpret a same terminology differently or assign a new name to an established and accepted terminology. The controlled vocabulary terms are called Zooclasses, abbreviated to Zoocs, and define concepts.

Experts from different domains are encouraged to submit proposals for new Zoocs to a Registrar. The Zooc may be very similar to existing Zoocs, however the differences need to be clear. After a successful peer review, including feedback from the proposers of any similar Zoocs, it is accepted into the Registry. Whilst centralised management was to be avoided there appeared no other way of limiting ambiguity and overlapping classification.

The requirement for simple and intuitive navigation through the classification resulted in a graphical representation. Zoocs are not presented singly, but are displayed together with narrower, broader, and related terms on a simple ontology called a Zooc Steering Diagram (ZSD), see Figure 2a to 2d.

It is the expert or group of experts that proposed the Zooc who assume the responsibility for the development of the corresponding ZSD: picking and placing Zoocs from the Registry to create a representation of their domain of expertise. The experts decide "what goes where" from their perspective.

The crowd is invited to navigate through the library of ZSDs to find and attribute Zoocs to the literature. The ZSD for the Pipe Zooc is shown in Figure 2a. The Pipe Zooc is called the Subject of the ZSD. Beneath the Subject are narrower terms that we call Types, in our example Pipes for liquids, gases and structure. Above the Subject are terms that are similar to the Subject. Collectively they are known as Sims. Sims can be unrelated terms that are similar to the Subject in a holistic manner: for example, the Tunnel and Trough Sims. Alternatively, Sims are similar due to their being a broader term, effectively a hypernym that we abbreviate to Hype. The Pipe Sim is such an example in the Pipe for liquids ZSD in Figure 2d. Polyhierarchies require multiple ZSDs and disambiguation.

A mouse pointer hovering over a Zooc will reveal the Zooc metadata and scope notes: see Figure 2b. Clicking on a Zooc loads its ZSD via a hyperlink: see Figures 2c to 2d. The display of related terms in a simple structure, of available scope notes and hyperlinking is we believe a simple and intuitive user interface.

The Sims are placed on the ZSD as a function of their similarity. So, the lower down the ZSD, and thus closer to the Subject they are, so the more similar they are to it. Whilst not shown, the sliding scale of similarity represents the weighting of a fuzzy classification. For the Pipe ZSD, shown in Figure 2a, a Tunnel is 40% similar to a Pipe, whilst a Trough is 20% similar. The Types on the other hand are all 100% Pipe.

For the purpose of search, we can choose to select the ZSD, now called a Zooc Similarity Diagram, rather than just the Subject. In our example we can expand a single Zooc query to include Pipes, Types of Pipe and Sims of Pipes, each with their related degree of similarity. This query expansion is analogous to the semantic query expansion of Tudhope and Binding (2008).

There is also an implicit expansion in terms of classification. As an example, the classification of a disclosure with the Tunnel Zooc will classify it as the Subject of the Tunnel ZSD, but also as 40% similar to a Pipe, given that it appears on the Pipe ZSD. If attributed to a disclosure it will be classified as many times as the selected Zooc appears on a ZSD, each time receiving the degree of similarity judged by the expert or experts in the field.

The ZSD is an ontology with the unique relationship of similarity. This is a key characteristic to being able to join up independently created ZSDs into what could be called a knowledge graph. The other characteristic is the fractal-like nature of the ZSD. Each Zooc that appears on a ZSD has its own ZSD "hidden" underneath it. These ZSDs in turn have Zoocs that have

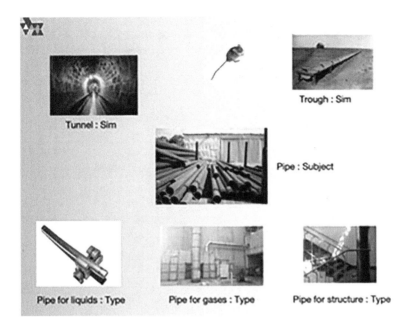

Figure 2a. Zooclass Steering Diagrams

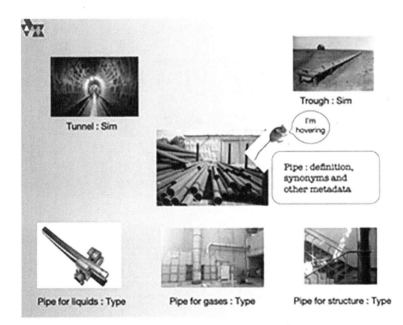

Figure 2b. Zooclass Steering Diagrams

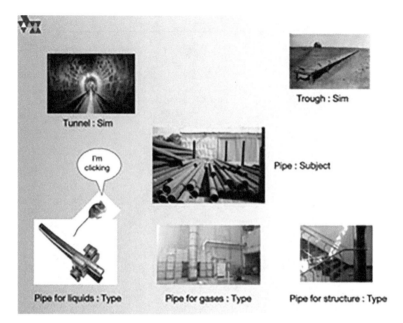

Figure 2c. Zooclass Steering Diagrams

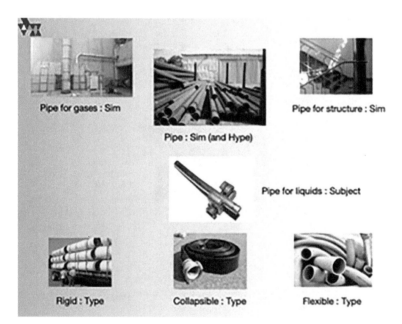

Figure 2d. Zooclass Steering Diagrams

ZSDs and so on down the levels, zooming down to reveal more detail akin to a Mandelbrot fractal.

As an example, the Tunnel Sim that appears on Figure 2a will have all the Tunnel Types and Tunnel Sims on it as decided by the Tunnel Experts. The Pipe Experts can leverage this work during a search. The Tunnel Sims are similar to something similar from the perspective of the Pipe. The unique relationship enables the overall similarity to be calculated with a simple algorithm as published by Absalom and Absalom (2012).

The fractal representation also serves to fill in the gaps of higher-level ZSDs. For example, if the Tunnel ZSD did not have the Trough Sim on it, but it included the Pipe Sim, the Trough Sim would be picked up on the first fractal level of the Tunnel ZSD via the Pipe Sim.

Going down the levels the algorithm will discover Zoocs that it has already seen before. The algorithm takes care of eliminating these repeat appearances: the first appearance with the fewest iterations of similarity is assumed to be the most accurate. Importantly the algorithm is run from each Subject Zooc enabling a knowledge graph to be developed with the perspective of each Subject Zooc expert.

Experts could map classification completed in different classification schemes onto their ZSDs. A perfect Zooc match could be placed alongside it, a close match placed in horizontal proximity as a function of the similarity.

As science and technology evolves so new Zoocs and ZSDs can be created. The newly created ZSDs can incorporate existing Zoocs and their ZSDs. Updating of existing ZSDs is distributed amongst the different experts.

Zoocs have living, zombie or dead status. Zombie status is attributed to a Zooc that has been superseded. Zombie classes cannot be attributed but are included for search for as long as all the literature that received the previously living class has not been reclassified. As an when reclassification is completed the zombie becomes a dead class maintained for information only.

Contributors only need to classify and search with Subject Zoocs, the cognitive task of estimating similarity with other concepts having been performed by the different domain experts.

To be granted a patent an application must be both new and non-obvious in relation to the prior art. A Subject Zooc and its Types can define all the different manifestations of a concept. This is useful for searching whether something is new. Non-obviousness excludes concepts that are too similar and is well served by the Sim design.

3.0 Selecting and combining multiple facets

Although originally intended as a search engine for patent and non-patent literature prior art, it was recognised at an early stage that the prior art is also a source of inspiration for new ideas. What facets would best support a search-and-innovation engine?

Problems and their solution are used to evaluate non-obviousness during patent prosecution. And many dormant patents have been described by Hartmann (2014) as solutions looking for a problem.

Cross-industry innovation applies known solutions to analogous problems to a different application or a different field of technology. Biomimetics seeks inspiration to unsolved problems from the natural world. Fortunately, analogy is similar to the similarity of the ZSD.

Elsewhere a problem may become a solution: the scanning tunnelling microscope's problem of attracting atoms from a surface under investigation provided the solution to the picking and placing of individual atoms was disclosed by Stroscio and Celotta (2004). The peelable adhesive that eventually found good use on 3M Post-it® notes is another good example disclosed by Hiskey (2011).

And so, an initial facet design was chosen including Solutions to a Problems in a context defined by an Application, Technology and Operation. As an example, a Fedora hat takes up too much room in the wardrobe. The Fedora hat is the Application or product. The Problem is the Fedora being too voluminous. The technologist called upon to investigate the Problem is an expert in textiles Technology. The Operation relates to the storage of the Fedora. The Solution is an improved memory effect in the textile of the hat that allows it to be collapsed flat for storage and then reformed for use, and this over the lifetime of the hat. This facet structure was abbreviated as ATOPS.

Zoocs from any-and-all ATOPS facets can be included in a query. The selected Zoocs are typically expanded to the Zooc ZSDs including the Subject, the Types, and the Sims, each of the latter with their similarity values. The fractal algorithm can continue the expansion as explained earlier. The resulting similarity lists are then matched with the ATOPS of the classified literature. The magnitude of any-and-all matches of the facets are combined as vectors in different facet dimensions. Combining the vectors provides a ranked list of holistically matched literature. Whilst the fuzzy values in all-and-any of the ZSDs remain somewhat arbitrary, the imprecision across multiple facets is of less concern: literature that matches in multiple facets is expected to rise high in the ranking.

Our first use case disclosed by Absalom, Absalom, and Hartmann (2012) considered the search for a stent, an artificial tube used in medicine to keep a body tube open as our Application. The Problem was stent thrombosis, where the stent becomes blocked. The Solution was a non-smooth surface as a lining.

We wondered if a simulated sharkskin lining to the stent would prevent material sticking to it in the same way a simulated sharkskin coating prevents the fouling of ship's hulls. Not having a corpus of classified literature our considerations remained hypothetical. In terms of the innovation engine, we imagined a situation where a catheter, a medical device similar to a stent as the Application, with the similar Problem of bacterial deposition, and with the Solution of a pimpled lithographed surface could have stimulated the non-smooth sharkskin lining, had it not existed. Whilst there would be no perfect match in either of the Application or Problem facets the catheter disclosure would rank highly. The pimpled surface would likely prove food for thought and alternative non-smooth surfaces considered.

A review of a mini pilot conducted in 2014 highlighted shortcomings of the basic ATOPS structure. We considered designs incorporating additional facet complexity. However, any theoretical increase in information-retrieval power from such increased complexity need take account of the increased cognitive burden on Registrars, Experts and Contributors. Would a theoretical increase in information-retrieval power be met in practice? Whilst simplicity is the ultimate sophistication,[1] everything should be made as simple as possible, but not simpler.[2]

We have considered alternative representations to enhance the design, including the provision of meronomy We have considered if-and-how artificial intelligence could bridge the gaps in ATOPS. We have also considered how to develop the wisdom of the crowd from multiple independent classifications of a same disclosure.

Recently we reviewed the stent example and realised that the Problem of bacterial deposition is better described as a cause of stent thrombosis than being similar to stent thrombosis. The rest of this paper will present a design for modelling such causality to enhance DZ.

4.0 Incorporating causality

Causality is complex. Studied in metaphysics as part of contemporary philosophy it was used by Robb (1911) to construct notions of time and space. More generally it sits at the intellectually demanding conjuncture of philosophy, physics, and mathematics, and has occupied many brilliant minds over thousands of years.

---

1   Attributed to Leonardo da Vinci.
2   Attributed to Albert Einstein.

Not unsurprisingly there are different schools of causality: regularity, probabilistic, counterfactual, mechanistic, and manipulationist.[3] All the schools require study. Such an investment is incompatible with even the most erudite and committed of crowds.

We require something simple and intuitive and would trade accuracy and academic rigour to meet these requirements. But it needs to support the search-and-innovation engine.

We looked to engineering. Root cause analysis (RCA) is used in both science and engineering to model the origins of problems and help find their solution. RCA uses causal graphs, where nodes representing causes and effect are joined by arrows to model their sequence in time. From a mathematician's perspective they are a directed acyclic graph (DAG).

That time travels uniquely in one direction with a cause necessarily preceding an effect makes the structure suitable for causality. Causal graphs and DAGs exclude directed cycles, where a cycle can travel forward in time but remain in a loop. In directed cycles an effect is a function of both the cause and of history. A simple example would be a waste bin that is push-to-open and push-to-close. Ignoring directed cycles maintains simplicity with minimal negative consequences for our design. RCA suggests the use of the Ishikawa diagram, a simple DAG, to brainstorm the root causes to a problem.

5.0 The Ishikawa Cause and Effect Diagram

Commonly called a fishbone diagram, Ishikawa (1968) designed diagrams to aid investigation into the causes of an effect or problem. An example is shown in Figure 3 below.[4] The first step in completing a diagram is the brainstorming of the different categories of cause, shown as the large bones along the fish's spine. Primary causes are then identified within each category and represented as smaller bones feeding into the large bones. The process is repeated identifying secondary causes that can cause the primary causes and be represented as even finer bones. The method terminates at the identification of the root causes of the problem. A repeated asking of why causes are produced, a technique formalised in Serrat (2017) as the 5 Whys, often accompanies the process. The Ishikawa diagram is simpler and more intuitive than causal graphs.

---

3   "Causality," Wikipedia, last edited February 1, 2022, https://en.wikipedia.org/wiki/Causality.

4   "Ishikawa diagrams," Wikipedia, last edited December 29, 2021, https://en.wikipedia.org/wiki/Ishikawa_diagram.

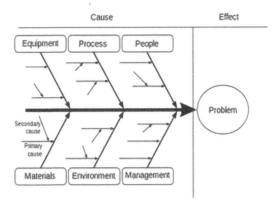

Figure 3. An Ishikawa Diagram

6.0 *The Modified Ishikawa Diagram (MID)*

Most patent and academic literature is not concerned with identifying and mitigating all the potential causes to a problem. Developing a single definitive diagram for a problem would likely be a major investment, even if done collaboratively. And different contexts may complicate and even compromise a resulting diagram.

It was therefore decided that the representation be unique to a particular disclosure, in a manner akin to document classification, or perhaps more akin to annotation.

Causes and effects can be desirable or nefast according to their context. Physics has principles. As an example, the Bernoulli *principle* relates an increased rate of horizontal flow with a reduction in pressure. The related Bernoulli *effect* underpins winged flight: very much a solution. It also produces the squat effect whereby the difference in speed of water passing underneath and aside a ship's hull in shallow water creates a downforce. This was a problem for the ship called the QE2 as disclosed by MAIB (1993), when it caused it to run aground.

From our perspective the causes and effects that end in a problem can all be viewed as problems. These can all be Zoocs in the Problem facet and as such the Bernoulli effect could be a Problem Zooc. In contravention with good thesaurus practice the Bernoulli effect can also be a Solution Zooc and be used to model Solutions: but this is the subject of a future paper.

A causal sequence is represented in a similar manner on the MID shown above. The end problem, shown here as PROBLEM at the base of the diagram, has a vertical timeline t=0. Direct problems A, B, C and D are placed

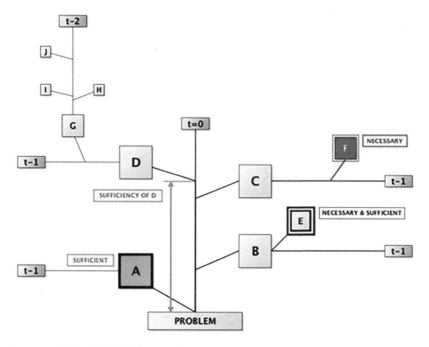

Figure 4. A Modified Ishikawa Diagram (MID)

alongside the t=0 timeline. They also appear at the base of their own t-1 timelines, indicating that they occur prior to the end problem. Intermediate problems, such as G in Figure 4, are placed alongside the t-1 timelines and have at their base a t-2 timeline. The process is repeated until all root problems are displayed, in our example problems A, E, F, H, I and J. It may well be the case that a disclosure only focuses on a single causal sequence.

In so far as they exist the problems should be represented as Problem Zoocs. If they do not exist, they should be referred to the Registrars for consideration.

Ishikawa diagrams have attracted criticism from Gregory (1992) for not including the logic of causation, and in particular the lack of provision for necessity and sufficiency conditions. We have included them on the MID.

Sufficiency is typically considered as a Boolean variable: a cause is sufficient or not sufficient to produce a subsequent cause. We interpret sufficiency as a continuous variable where the sufficiency of a problem, effectively its ability to produce the subsequent problem, is represented by the height from the base of the subsequent problem's timeline. Direct Problem A, shown in green at the base of the t=0 timeline is sufficient, meaning that every time it occurs it will produce the problem. Problem D is only some-times sufficient, perhaps 3 times in 10, against 4 times in 10 for Problem C and 8 times in 10 for Problem B. Insufficient problems, those that cannot produce the subsequent problem, as not shown on the diagram as they do not represent a problem.

It is important to distinguish the sufficiency of a problem from the like-lihood it will arise in the first place. There is no graphical representation of such likelihood on the MID. Unlike engineering or science modelling we are not seeking to quantify a problem, to determine a failure rate or calcu-late the risk or severity of an outcome. Existing frameworks exist to serve such purposes. Fault Tree Analysis (FTA) uses Boolean logic and statistical probabilities of component failures to model how systems fail, and to de-sign mitigation strategies. Failure Mode and Effects Analysis (FMEA) and its derivative Failure Mode, Effects and Criticality Analysis are related and popular frameworks.

That said, a MID could eventually be annotated with a simplified likeli-hood of occurrence, such as FMEA's extremely unlikely, remote, occasion-al, reasonably possible, and frequent scale. Alternatively, a colour-coded log scale could be developed, or the length of the connection between the sub-sequent problem's timeline and the causal problem could be used. However, likelihood is often represented by a statistical distribution.

Quantifying sufficiency is also difficult given that there may be many variables. Braking hard in a car can result in the wheels locking if the car does not have ABS brakes. Braking harder will increase the sufficiency. The sufficiency will also be higher on a wet road than on a dry road, and even light braking may be fully sufficient on icy roads. As an alternative example imagine a glass of water on a table in a train. It could be sent flying if the train hits an obstacle on the line. The same could happen from a large jolt of the train due to a damaged track. A worn track could produce a series of jolts with the same result. And a poor-quality track could produce a vibration to do the same. Such single, multiple, and constant problems of differing mag-nitude are easier to define than they are to quantify. That said, combining likelihood with an empirical quantification of sufficiency could produce a simplistic ranking of how big a problem the different causal sequences are.

A necessary cause is shown in red in Problem F on Figure 4. Problem C can only happen in response to Problem F. Of note is that Problem F is alone on C's t-1 timeline. Identifying necessary causes is useful in that mitigating them removes the subsequent problem. More exactly the necessary problem needs to be prevented rather than cured. Of note is that Problem E is both necessary and sufficient.

The MID serves to link related problems for the purposes of search and innovation. Annotating the different Zooc Problems with related ATOS

Zoocs would appear a worthwhile endeavour. The basic ATOS, and especially the Operation facet, requires sub-faceting to facilitate this. Sub-facets are required to identify the different stages of a product lifecycle and well as operations *per se*. A work in progress.

The MID can supplement search with ZSDs. A search query, with all-or-any ATOPS facets could proceed using ZSDs and the fractal algorithm as per the original design, whilst at the same time searching for the query Zoocs in the MIDs, including the MID Zoocs proper and ATOPS annotations. As an example, the disclosure with the MID of Figure 4, could match the Problem Zoc of the query with Problem F, where Problem F had not been attributed as a classification in the original manner. Related solutions to F could appear in the text of the Problem F disclosure and/or be included as ATOPS annotations to the MID.

The MID can be transposed into an equation. Direct problems A, B, C and D can all produce the end problem. Described in Boolean logic they have an OR relationship that is represented by a + symbol. A causal sequence can be represented as a sequence from end problem back to the root problem separated with commas. The sequence goes back in time from left to right. We have chosen to represent sufficiency with ^ and necessary with ! symbols.

$$PROBLEM = \{A^\wedge\} + \{B, E!^\wedge\} + \{C, F!\} + \{D, G, H+I+J\}$$

We are currently studying methods to match MID equations that have been developed for different documents. This could leverage algorithms from natural language processing such as spell-checkers. However, we have started with disassembling equations into triples of causal problem, effect problem and the sequence distance between them.

7.0 *Contributory causes*

Contributory causality can model situations where multiple causes can together cause an effect. Mackie (1974) proposed the INUS model where contributory causes are Insufficient but Non-redundant parts of a condition, which is itself, Unnecessary but Sufficient for the occurrence of an effect.

An example in Wikipedia[5] describes a short circuit, the proximity of flammable material, and the absence of firefighters being INUS conditions for a house burning down.

We prefer that sufficiency be represented as either sometimes sufficient or fully sufficient as on the MID. Contributory causes also need be represented

---

5   "Causality," Wikipedia, last edited February 1, 2022, https://en.wikipedia.org/wiki/Causality.

as individually insufficient. We see the commonality between contributory causation and fuzzy logic and are presently investigating methods to integrate it with MID diagrams and MID equations. Our goal is to develop an all-encompassing MID-logic.

8.0 Conclusions

We believe that the work presented here lays the foundations for further study. A crowd will eventually be necessary to classify a corpus of literature to be able to test the ideas. In the shorter term there remains much to do: with the facet complexity, with developing the wisdom of the crowd from multiple independent classifications, and with MID-logic. The use of artificial intelligence to process the causal relationships of the MIDs, grafting them together to build larger and more complex models remains a very distant goal. We remain open to ideas and collaboration.

References

Absalom, Arthur and Geoffrey Absalom. 2012. "Durham Zoo: prior art and solution search." Durham Zoo. Last accessed February 10, 2022, http://durhamzoo.org/wp-content/uploads/2012/11/Durham-Zoo1.pdf.

Absalom, Arthur, Geoffrey Absalom, and Dap Hartmann, D. 2012. "A collaborative classification-based search engine for prior art and solution search." In *2012 IEEE Sixth International Conference on Semantic Computing (ICSC 2012) 19-21 September 2012 Palermo, Italy*. Institute of Electrical and Electronics Engineers (IEEE), 87-92. https://doi:10.1109/ICSC.2012.30.

"Causality." Wikipedia. Last edited February 1, 2022. https://en.wikipedia.org/wiki/Causality.

Gregory, Frank. 1992. "Cause, Effect, Efficiency & Soft Systems Models. Warwick Business School Research Paper No. 42". *Journal of the Operational Research Society* 44, no. 4:333-44. https://doi.org/10.1057/jors.1993.63.

Hartmann, Dap. 2014. "Turning Technology into Business Using University Patents." *Technology Innovation Management Review* 4, no. 12:37-43. https://doi:10.22215/timreview/856.

Hiskey, Daven. 2011. "Post-it notes were invented by accident." *TodayIFoundOut*. Last accessed February 10, 2022. http://www.todayifoundout.com/index.php/2011/11/post-it-notes-were-invented-by-accident/.

"Ishikawa diagrams." Wikipedia. Last edited December 29, 2021. https://en.wikipedia.org/wiki/Ishikawa_diagram.

Ishikawa, Kaoru. 1968. *Guide to Quality Control*. Tokyo, Japan: Asian Productivity Organization.

Mackie, John Leslie. 1974. *The Cement of the Universe: A Study of Causation*. Oxford, UK: Clarendon Press.

Marine Accident Investigation Branch (MAIB). 1993. "Queen Elizabeth 2 - grounding of a passenger vessel on 7 August 1992".

Robb, Alfred A. 1911. *Optical Geometry of Motion*. Cambridge, UK: W. Heffer.

Serrat, Olivier. 2017. "The Five Whys Technique," *Knowledge Solutions*, 307-10. https://doi.org/10.1007/978-981-10-0983-9_32.

Stroscio, Joseph A. and Celotta Robert J. 2004. "Controlling the Dynamics of a Single Atom in Lateral Atom Manipulation." *Science* 306, no. 5694:242-47. doi:10.1126/science.1102370.

Tudhope, Douglas and Ceri Binding. 2008. "Faceted Thesauri." *Axiomathes* 18, no. 2:211-22. https://doi:10.1007/s10516-008-9031-6.

Licenses for the use of images (lack of place with the images themselves)

Figures 2a to 2d: Mouse pointer: George Shuklin, CC BY-SA 1.0 <https://creative commons.org/licenses/by-sa/1.0>, via Wikimedia Commons

Tunnel: Thomas Bresson, CC BY 2.0 <https://creativecommons.org/licenses/by/2.0>, via Wikimedia Commons

Pipe for liquids: Tommy Halvarsson, CC BY-SA 3.0 <http://creativecommons.org/licenses/by-sa/3.0/>, via Wikimedia Commons

Pipe: Paul Goyette, CC BY-SA 2.0 <https://creativecommons.org/licenses/by-sa/2.0>, via Wikimedia Commons

Pipe for gases: Combustion2016, CC BY-SA 4.0 <https://creative commons.org/licenses/by-sa/4.0>, via Wikimedia Commons

Pipe for structure: Holger.Ellgaard, CC BY-SA 3.0 <https://creativecommons.org/licenses/by-sa/3.0>, via Wikimedia Commons

Collapsible: Thiemo Schuff, CC BY-SA 3.0 <https://creativecommons.org/licenses/by-sa/3.0>, via Wikimedia Commons

Flexible: Paul Goyette, CC BY-SA 2.0 <https://creativecommons.org/licenses/by-sa/2.0>, via Wikimedia Commons

Figure 3:   Ishikawa fishbone-type cause-and-effect diagram: FabianLange @ de.wikipedia <https://en.wikipedia.org/wiki/GNU_Free_Docu mentation_License>

The FITS file format for the long-term preservation of digital objects in the Environment and Earth Observation domain

Stefano Allegrezza
University of Bologna, Italy

Abstract

One of the most critical problems in the long-term preservation of digital objects in any domain is the rapid obsolescence of file formats that become outdated in a short time and therefore no longer readable. This problem also affects the field of environmental and earth observation, where it is important that the data collected remain accessible for many years to come. For this purpose, in recent years the attention has been focused on an extremely versatile file format created for the astronomy and astrophysics domain but then spread to other sectors as well: the Flexible Image Transport System (FITS). What exactly is it? What are the characteristics of the FITS format that made it so interesting to be a candidate for adoption in the Environment and Earth Observation (EO) field? Is a format that was created over forty years ago still appropriate for long-term preservation? This paper aims at answering these questions by starting from an analysis of the FITS file format and highlighting its features in order to understand if it is appropriate for the archiving and preservation of data collected by many scientific projects in the EO domain.

1.0 Introduction

One of the most worrying problems in the long-term preservation of digital objects in any domain is the rapid obsolescence of file formats that become outdated in a short time and therefore no longer readable. This problem also affects the Environment and Earth Observation (EO) domain, where it is important that the data collected remain accessible for many years to come.

> "Earth Observation data are unique by nature and are fundamental for the monitoring of our environment and planet and of its changes. They are considered as humankind assets and as such need to be preserved without time constrains and kept accessible together with all the information and knowledge needed to understand and use them in future" (Albani 2012).

There are strong motivations for preserving Environment and Earth Observation data (National Research Council 1995). For instance, many observations about the natural world are records of events that will never be repeated exactly. Examples include observations of an atmospheric storm, a deep ocean current, a volcanic eruption, and the energy emitted by a supernova. Once lost, such records can never be replaced. Observed data provide a baseline for determining rates of change and for computing the frequency of occurrence of unusual events. They specify the observed envelope of variability. The longer the record, the greater our confidence in the conclusions we can draw from it. Data records may have more than one life and can be reused. As scientific ideas advance, new concepts may emerge – in the same or entirely different disciplines – from the study of observations that led earlier to different kinds of insights. New computing technologies for storing and analyzing data enhance the possibilities for finding or verifying new perspectives through reanalysis of existing data records. Thus, the relative importance of data, both current and historical, can change dramatically, often in entirely unanticipated directions. Finally, the substantial investments made to acquire data records justify their preservation: because we cannot predict which data will yield the most scientific benefit in years ahead, the data we discard today may be the data that would have been extremely valuable tomorrow.

Long-term preservation of EO data and of the ability to discover, access and process them is a fundamental issue and a major challenge at programmatic, technological and operational levels (Albani and Giaretta 2009). In order to create and administer archives that serve current and future generations of users adequately, archive owners require explicit and precise requirements from data users (Molch et al. 2012). Furthermore, curating and preserving EO data is important to justify funding for long-term preservation in today's challenging economic climate, archives must be able to present a convincing case for the value of the data in the long term (Conway et al. 2014).

Because of the rapid technological change, the long-term storage of EO data comes with its own set of risks and difficulties. The corruption of the bitstream, as well as existing hardware and operating environments, are all dangers that render data inaccessible on a physical and logical level. However, insufficient data description, inability to discover data, and service compatibility might all limit re-use (Albani 2012). There is another major difficulty in the EO area that makes long-term preservation of obtained data particularly difficult and contributes to the high operational and maintenance expenses of long-term archiving: the excessive proliferation of diverse and heterogeneous data formats. This is caused by mainly three reasons:

"the lack of an agreed standard in the EO community, reason for which the formats tended to be specific for the sensors each mission carried on board; legacies from old ground segments architectures, which tended not to reuse elements previously developed; the non-mature status, until recently, of the information technologies and standards used to describe and package the data, preventing the creation of a unique format able to satisfy at the same time the requirements for the long-term preservation of the data and their handling in the processing centres" (Pinna and Mbaye 2014).

To overcome this problem digital preservation specialists, knowing that digital preservation requires their continuous migration to other formats over time, suggest reducing the number of formats to manage; and, better, using file formats that give the broadest guarantees of long-term sustainability.

For this purpose, in recent years the attention has been focused on an extremely versatile file format that was created for the astronomy and astrophysics domain but that is now spreading to other fields: the Flexible Image Transport System (FITS). This paper aims at evaluating the FITS file format in order to understand if it is suitable for archiving and preserving data collected by many scientific projects in the EO domain – such as the ERA-PLANET project that has generated a substantial amount of data and knowledge on different aspects related to the environmental quality and sustainability – and to make it easier to share them with stakeholders and policy makers and to support decision making.

2.0 State of the art

Over the past fifteen years, leading EO organizations have launched projects to define the content to be preserved over time (Ramapriyan et al. 2017). For example, the European Space Agency (ESA) formed a Long Term Data Preservation (LTDP) Working Group in 2007 with the aim of defining and promoting a coordinated approach to the preservation of the European EO data assets. One of the outputs of this working group was the "Earth Observation Preserved Data Set Content" (EO PDSC), a document providing guidance to data holders on preservation. There have been several versions of this document, the latest having been published in 2012 (ESA and CEOS/WGISS 2015; ESA 2105). In late 2011 NASA developed its Earth Science Preservation Content Specification (PCS) (NASA 2011) and has been using it as a requirement for its new missions. For missions that had been in progress or completed before the PCS was developed, it is used as a checklist to capture and preserve as many relevant content elements as possible based on best efforts. Other projects have also addressed these issues but the issue of electronic formats has only been partially addressed.

A very large number of file formats and solutions have been proposed in the EO domain. The major players and many research groups have proposed their own file format for data archiving, sometimes without any care for long-term preservation, although in the last 15 years several projects were funded to address the long-term preservation issue of scientific data in general and a dedicated ISO standard has been developed, ISO 19165-2:2020, which aims to provide details about content describing the provenance and context specific to data from missions that observe the Earth using space-borne, airborne or in situ instruments (ISO 2020). Some of the most adopted file formats include:

– HDF (Hierarchical Data Format), a set of file formats (HDF4, HDF5, HDF-EOS 5) designed to store and organize large amounts of data. Originally developed at the National Center for Supercomputing Applications (NCSA), HDF is currently supported by the HDF Group, a non-profit corporation whose mission is to ensure continued development of HDF5 technologies and the continued accessibility of data stored in HDF (https://www.hdfgroup.org);
– CDF (Common Data Format), a file format (together with a library, and a toolkit) that was developed by the National Space Science Data Center (NSSDC) at NASA starting in 1985;[1]
– NetCDF (Network Common Data Form), a self-describing, machine-independent file format and a set of software libraries that support the creation, access, and sharing of array-oriented scientific data. It is commonly used in climatology, meteorology and oceanography applications (e.g., weather forecasting, climate change) and GIS applications. The project is hosted by the Unidata program at the University Corporation for Atmospheric Research (UCAR);[2]
– GeoTIFF (Tagged Information File Format), an extension of TIFF that includes georeferencing or geocoding information embedded within a TIFF file (such as latitude, longitude, map projection, coordinate systems, ellipsoids, and datums) so an image can be positioned correctly on maps of the Earth;
– SAFE (Standard Archive Format for Europe), a file format designed to act as a common format for archiving and conveying data within ESA Earth Observation archiving facilities. SAFE has been designed to be used in an archive system compliant with the Open Archival Information System

---

1 NASA, "Space Physics Data Facility," last accessed December 13, 2021, https://cdf.gsfc.nasa.gov/html/FAQ.html.
2 Unidata, "Network Common data Form (NetCDF)," last accessed December 13, 2021, https://www.unidata.ucar.edu/software/netcdf.

(OAIS) standard. SENTINEL-SAFE format wraps a folder containing image data in a binary data format and product metadata in XML;[3]

- ASDF (Advanced Scientific Data Format) is a proposed replacement to the FITS standard for astronomical images and other astronomical data. The metadata is contained in a YAML (Human-readable data serialization format) header followed by binary or ASCII data;[4]

- GRIB (GRIdded Binary or General Regularly-distributed Information in Binary form), a concise data format commonly used in meteorology to store historical and forecast weather data. It is standardized by the World Meteorological Organization's Commission;[5]

- PP-format (Post Processing Format), a proprietary file format for meteorological data developed by the Met Office, the United Kingdom's national weather service;

- GFS (Global Forecast System), the format used by a global numerical weather prediction system containing a global computer model and variational analysis run by the United States' National Weather Service (NWS);[6]

- CEOS (Committee on Earth Observation Satellites), a standard format published in 1988, used for radar data and originally expected to be used with tape media. The format does not specify a naming convention (https://ceos.org).

There are also many other file formats, such as CCRS, EOSAT, AVSAR, AIR-SAR, DORADE, Cf Radial, LGSOWG, UNIVERSAL Format (UF), FORAY NetCDF Format, OGC KML, Fast-L7A and the IWRF time series format.

From the point of view of long-term preservation, the fact that in the field of EO there are so many and diverse file formats is a serious problem. In fact, you have to consider that you need to manage all these formats (e.g., you need to migrate them over time), and managing a few formats – or better just one – is definitely easier than managing several dozens of file formats.

To overcome this problem digital preservation specialists suggest to reduce the number of formats to manage and to prefer formats that give the broadest guarantees of long-term preservation. The perfect solution would be a versatile file format (e.g. we can use it in different application areas), and

---

3   ESA, "SAFE," last accessed December 13, 2021, https://earth.esa.int/SAFE/index.html.

4   STSI, "ASDF Standard," last accessed December 13, 2021, https://asdf-standard.read thedocs.io/en/1.6.0.

5   COSMO, "GRIB Data Format used for tge COSMO-Model System Features of GRIB," last accessed December 13, 2021, http://www.cosmo-model.org/content/ model/documentation/grib/grib_ features.htm.

6   NCEI, "Global Forecast System," last accessed December 13, 2021, https://www.ncei. noaa.gov/products/weather-climate-models/global-forecast.

a file format that do not need to be migrated for at least a few decades. In this regard, in recent years the attention has been focused on the Flexible Image Transport System (FITS).

3.0 The Flexible Image Transport System

FITS is a widely used file format for storing images and data in astronomy and astrophysics (Wells and Greisen 1979; Greisen et al. 1980; Wells et al. 1981; Greisen and Harten 1981; Hanisch et al. 2001). It was created in the early 70s to solve the problem of the great variety of electronic formats existing in that period. Since every institution in the late Seventies had its own way to keep the collected data, it became necessary to establish a standard in order to avoid the waste of time and resources usually spent each time to develop a customized software to convert the received data to the format used at the home institution and so forth.[7]

The year of birth of the format is dated 1979, but the official year of birth of the format is considered 1981, when the specifications of the file format where published in the "Astronomy and Astrophysics Supplement Series" journal thanks to an article written by Donald C. Wells, Eric W. Greisen, and Ronald H. Harten (Wells et al. 1981). With the help of another author, Preben Grosbol (Grosbol 1988), during a seven year period of time (1981-1988) they published three more papers in the same journal to better describe the specifications of the format: all together they are known as the 'Four FITS papers' (NASA 2014).

In 1982 the International Astronomical Union (IAU), a non-profit international association of scientists of the space and astronomical sector, formally adopted it and in a short time the FITS file format became a de facto standard for the interchange of data. However, the first official version of the FITS Standard was drafted by the NASA Office of Standards and Technology (NOST) at the Goddard Space Flight Center, and formally approved by the IAU FITS Working Group (IAU-FWG)[8] only in June 1993, under the name NOST 100-1.0. The second official version, called NOST 100-2.0, dates back

---

7   The main motivation for the introduction of the format at the time stemmed from the fact that the majority of scientific organizations operating in the field of astronomy used for storing scientific data, proprietary file formats and not supported by the software in use by the other organizations. It was felt, therefore, the need to overcome the situation of insufficient interoperability then widespread thanks to a common format that would enable the interchange of data.

8   The IAU FITS Working Group consists of the following regional committees: the European FITS Committee; the Japanese FITS Committee; the American Astronomical Society FITS Committee; the Australian/New Zealand/Pacific FITS Committee.

to March of 1999. This version was published in the "Astronomy and Astrophysics" journal in September of 2001. Seven years later a new version, numbered 3.0, was officially approved by IAU-FWG in July 2008 and remained valid until 2016, when the fourth version of the FITS Standard was made public and is currently in force.

Table 1 lists the various versions of the format that have occurred since its birth. It is worth noticing that the file format has been very stable throughout the years, since it has gone through only four major versions in about forty years.

Version	Name	Date	Notes
	-	1979	FITS birth
	-	1981	Official FITS birth
	-	1982	FITS adopted by IAU
0.1	NOST 100-0.1	December 1990	1st draft
0.2	NOST 100-0.2	June 1991	2st draft
0.3	NOST 100-0.3	December 1991	3rd draft
1.0	**NOST 100-1.0**	**June 1993**	**NOST Standard (1st version)**
1.1	NOST 100-1.1	September 1995	Some amendments
1.2	NOST 100-1.2	April 1998	Some minor amendments
2.0	**NOST 100-2.0**	**March 1999**	**NOST standard (2nd version)**
2.1	IAUFWG 2.1	April 2005	IAUFWG standard
2.1b	IAUFWG 2.1b	December 2005	IAUFWG standard (64 bit integers support added)
3.0	**IAUFWG 3.0**	**July 2008**	**IAU-FWG standard (3rd version)**
4.0	**IAUFWG 4.0**	**July 2016**	**IAU-FWG standard (4th version)**

Table 1. The evolution of FITS

The FITS format specifications are described in a document called "The FITS Standard" maintained by NASA.[9] The latest version, consists of a text of about 75 pages divided into sections which fully covers the whole range of the format's details, such as the file organization, headers, keywords, data

---

9   NASA, "FITS Standard Document," last accessed December 13, 2021, https://fits.gsfc. nasa.gov/fits_standard.html.

representation, standard extensions, the world coordinate system, etc. Currently the IAU-FWG maintains the format adding updates and improvements.

Although FITS was developed by astronomers as the standard file format for the interchange of images between hardware platforms and software applications that did not share a common file format, currently it has become the standard format for archiving and preservation of astronomical images of many scientific organizations, such as the images captured from radio telescopes and other astronomical equipment or those returned by orbital satellites, from the spacecraft and from the planetary probes in the course of their missions (for example, NASA uses it to store the images of the space missions).

3.1 How FITS organizes data

A FITS file consists of a sequence of one or more blocks, each of which is referred to as HDU ("Header and Data Unit"): the first is referred to as "Primary HDU" and the following are referred to as "Extension HDU" (NASA/ GSFC 1997). Each HDU in turn consists of two sections: a header (called "Header Unit" or simply "Header") and an optional section containing data (called "Data Unit" or "Data Array"). In the case of the "Primary HDU" the header takes the name of "Primary Header Unit" and the data section takes the name of "Primary Data Unit" (or also "Primary Data Array"), while, in the case of an "Extension HDU" the two sections are called "Extension Header Unit" and "Extension Data Unit" (or "Extension Date Array"), respectively (see Figure 1). The simplest FITS file contains a header and a single data stream. The simplest file in FITS format consists of a single Primary HDU.

FITS can contain data arrays that range in size from 1 to 999. In the case of size "2" it is called the matrix and the data are typically two-dimensional digital representation of an image in raster format.

The format structure is flexible and allows to add any number of Extension HDU depending on the amount of information you need to store. Currently the format specifications include three standard types of Extension HDU (NASA 2014):

- *Image Extension HDU*, used to store arrays of pixels in size between 0 and 999 (exactly the same type of data stored in the Primary HDU). In this case the Header Unit begins with the string XTension = 'IMAGE').
- *ASCII Table Extension HDU*, used to store information in a table using data encoded in ASCII. ASCII tables are generally less efficient than binary tables, but can be constructed to be easily read by humans, and are

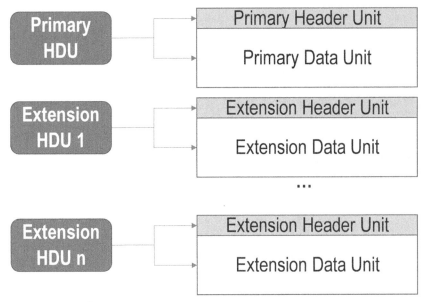

Figure 1. FITS file structure

capable of storing numeric information with arbitrary size and precision. In this case (header begins with XTENSION = 'TABLE').

– *Binary Table Extension HDU*, used to store information in tabular form using a binary representation. In this case the Header Unit begins with the string XTension = 'BINTABLE'. Each cell of the table can be a matrix, but the size of this array must be constant within a column. Thanks to this type of extension it is possible to use the FITS format to store, for example, entire relational databases.[10]

Each Header Unit consists of one or more records, known as keyword records (or even "card image"), each of which has a length of 80 characters and is encoded using the set of characters between the restricted ASCII character "SP" and the ASCII character "~" (126 decimal or hex 7E) (see Figure 2).

For historical reasons, the size of each Header or Data Unit must be an exact multiple of 2,880 bytes (corresponding to 36 alphanumeric strings with a length of 80 bytes each). If necessary, any unused space at the end of a Header or Data Unit can be completed with fill characters up to this dimen-

---

10   In addition to the structures mentioned above, there is another type of HDU called "Random Group" that is used almost exclusively for applications of radio interferometry, and whose use is not recommended for other types of application outside of it.

```
SIMPLE   =                    T        / file is compliant to FITS standard
BITPIX   =                   32        / Number of bits per pixel
NAXIS    =                    2        / Number of axis
NAXIS1   =                 1000        / lenght of data axis 1
NAXIS2   =                 1500        / lenght of data axis 2
DATE     = '2021/07/14'               / creation date
ORIGIN   = 'ESA'                      / FITS file originator
AUTHOR   = 'John Brown'               / creator
OBJECT   = 'Temperature Measurements' / name of object
HISTORY    'First version'
COMMENT    'Example of Header Unit'
END
?,,»?Šó?[1]Û>Ðó¢?h?qß<?•^?-É?¯Ũ×>†$¼Û<j=š  ?Ũ ÇÔ?>z>Ò"Ô>Û¼æ>*-O?%%¤?)£>šM?+?¯?\?7"=?#;Ç?
[1]i>|?t<ê?g!ß?j^Â?q?Bž~?\[Ž?'´¤?¤½Þ? @[?±I=?o|>?<^?è_4?Ý®à@?PK?<ƒ<?\>x?KDÏ?zØ;?Ã##?.
‹?.3\?°âP@ht@E£@RÉ^@y>O@§¦,@¦@ÐµÔ@òhmA[1]éq@ÛÉ•@Û¼sA7ôôAzb€AfxFAV9LAD[¿Aó1A,†A2I›A4ó'
A}(A†¬›A§TÚAq³AHjZAMe+A}¬AŠóAž[1]$A¯¶óA³/AÂAÂÑúÂÂæ•|B(LÏB1B)à£BAôTÇAY,úAKE¦A.0¢Aß,Aó
@â¤Ï@àZ@¨Ãš?¢?L™š>_u'?;dâ>®á†>RT>âr™?Ké?D"?°¢>ß3?V™%?Q1?€¦†?ë[>mëw?Ũ <1?¬>X?v?bœÔ>à3?¯
?;½?is?Ž»?°ê³?'+?›m?gt1>ÉÉS><pw?GXt?€#4? #;?°à7?¿8?1X0?DRç?Œ¸b?ÛkV?ÈuÔ?¬(™?Šùp?œÝ¬?ê
[1]Ò?äy:?Î~¦?¢¿€?Êî¦?'dÐ?Áîƒ?¬¦p@Ý@5û?èòsdjklaakkal
[…]
Ê±êÔà@Q£@'Œ@†vi@šÉÝ@¦Òž@¨Ž@œœ§9@±ÀÝ@è"=@èÉ@p´@áq|AÀA'¬œA†$ÀAÉFÀA¿†ôAš¨AÈA^
BA²¿oAΆAñ:IB $B[1]…B™7B„ôB+>iB\^B¦B{Û3BpBÙ[1]A¯ ,A^dKAd°AÑoAà9@óµ¸A¼Ý@Ý;!?ß?Þ°Ž?H
j…?q³È>Ð©?šÐÂ?Xaá@ó>àü¨?¨^Â?Àü>h³Š^>Û^Y>£¶Â?Ò>Ð5»?_„?^oÂ?½½?Y,?'À>ò¨¨?)Ò?¦j*?«j2?¢
q?fàé?¥?žŒÒ?A^-?AÂ?Û×ò?^£G?®®û?›V?[1]?›ûA@¹€?ÏùéŸ?§?Œál?Yjó?³zW>ëL:=ƒK?kJç?£ŽS?Ø ?¢
?\?†°1?"RΦ?»•Á?Ï½"?µ¼¤?´0 4?"ih?¸RD?† Û¨?…v¦?Šêá?ÒÏd?Óqf?Ôâ$?ôôÒ?¿† ?£m?âR¿½?Ð1?<¸<?¯¼p?¸
ƒ•?±tO@ Ý>Î\¯?{óÎ???90Û?1<?¼o0? $ó>-}?«6¶?µçÆ?‡"?³×^@1¬m>,#½?¸Î¦>Y¦G?°úà?ÀÒ?IÆt?„
ó?P®€?<²?é#?W?P´{>Ûô7>«Ði?¸Ýí?ê?n+µ>½c3>¦¦¯>èW¯>²êw?¸³Ò?Y³>xB?{íG?qy>ù|>?H…t>à8|¼É÷¨
¿F[>:¿~?•yt?Ëxò?Kôž?!O?tâN?8Ð¼¿Ç44y+8?Œ
```

Figure 2. A simple FITS file

sion: in the case of a Header Unit the ASCII character "SP" (32 decimal or 20 hex) should be used, while in the case of the Data Unit the ASCII "NUL" (0 decimal or 00 hex) will be used.

Both the Header and the Data Unit consist of logic blocks that need to be the size of 2,880 bytes or multiples thereof, whereby if the last logical block of one unit does not reach that dimension many characters (the AS-CII character "SP" in the case of the header unit or the ASCII "NUL" in the case of the Data Unit) filling up to the exact size of 2,880 bytes or a multiple thereof must be added. In turn, each logic block contains 36 records, called keyword records, each 80 bytes in size, possibly supplemented with fill characters to reach that length. As can be seen, this is an extremely linear logical organization.

3.2 The strengths of the FITS file format

Currently FITS is the most used file format for scientific data in astronomy and is also widespread in the field of astrophysics.

One of the strengths of the format is certainly its extreme simplicity: the structure of a file encoded according to the FITS format is, in fact, extremely straightforward. Essentially a file in FITS format consists of two main components, the Header unit containing the metadata and the Data unit containing the actual data. This organization can be repeated indefinitely (there is no limit to the number of HDU Extension that can be added), making the format scalable. FITS has been designed specifically for scientific data and therefore provides the possibility to use the metadata to completely describe the object represented and the context of production.

Another key for its success is its flexibility. Despite the letter "I" in the name of the format (Flexible "I" mage Transport System) is the initial of the word "Image", actually FITS not only allows the exchange and storage of images but is of much more general use.[11] In fact, thanks to the Extension HDU mechanism, it is possible to store digital objects of various kinds (not only images but also spectra, photon lists, data cubes, multi-structured data such as multi-table databases etc.). There are no limits to the numbers of Extension HDU you can add to a FITS file (so there are no limits to the amount of data you can put in a FITS file). In addition, each of the extensions may contain different types of data objects. So, for example, it is possible to store x-ray and infrared exposures in the same file. As the term 'flexible' suggests, it is a very flexible file format for transporting, analyzing, archiving and preserving scientific data files (Greisen et al. 1981). In this sense FITS may be considered a general-purpose storage format for data, and thus can be a good solution also in the EO field.

Although FITS is the most commonly used digital file format in the astronomical and astrophysics field, in recent years, thanks to its flexibility, it has also been used in other fields, such as the cultural heritage sector. For example, the Vatican Apostolic Library adopted the FITS file format as the default format for storing and preserving images from its massive manuscript digitization project of the 80,000 most important manuscripts in the world (Allegrezza 2011; Ammenti 2012).

It was the first institution that decided to use the FITS file format for something different from scientific purposes: the preservation of cultural heritage (Manoni et al. 2018). Given the great success obtained with the digitization and permanent storage of several thousands of manuscripts, as well as the usage of state-of-art features in terms of online accessibility by adopting the International Image Interoperability Framework (IIIF), this seems to be the way forward for every serious cultural institution aiming to follow the best practices of the Vatican Apostolic Library. FITS also begins to be used in the health sector to store and preserve data from imaging systems (e.g. computerized tomography in nuclear medicine).[12]

---

11   In the specifications of the format there is no constraint which limits its use to images, therefore it may well be considered a general-purpose storage format for data.

12   The FITS format is widely used as an image format in the field of astronomy and astrophysics, but its use in digitization projects was certainly not envisioned by its

Another important factor for the widespread adoption of FITS is the ease of writing the code for its use, given by the fact that all the documentation has always been publicly available. As a result, a large amount of software has been developed throughout time, primarily for two purposes: displaying material and converting to and from the FITS data format. The importance of this feature cannot be overstated, because one of the main causes of data obsolescence is the direct link between a digital format and the private corporation that created it, which, once off the market, leaves future users without the necessary tools to run it.

4.0 Evaluation of FITS as a digital preservation file format

In order to evaluate this format against digital preservation requirements, it is useful to consider a series of requirements a file format should have to be considered a "preservation format". In particular, we will evaluate FITS file format against the sustainability factors proposed by the U.S. Library of Congress (Library of Congress 2021), with a few adjustments:

- *non-proprietary*: FITS format is managed by the international scientific community (in particular, the astronomers and astrophysics community) and, as such, it is not proprietary. The retention of specifications is currently borne by the IAU FITS Working Group;
- *standardization*: FITS is a standard format recognized since 1993; in addition it is an open standard, being a format that has both the properties of standardization and openness;
- *disclosure*: FITS is an open format, being based on a specification called "The FITS Standard" that is freely available; there are no royalties (and never will be) to pay for the use of the format, so anyone can create applications that can create and manage FITS files;
- *documentation*: FITS format is fully documented in its specification, "The FITS Standard";
- *adoption*: The FITS format is currently the most commonly used format in astronomy; there are also numerous applications for creating and viewing files in FITS format and for their conversion from and to other image formats;

---

creators. This is probably due to the fact that among the many unresolved problems involved in preserving cultural heritage in digital format (fragile storage media, hardware and software obsolescence, obsolescence of electronic formats, etc.) the FITS format seems to promise a solution to at least one of the main obstacles encountered in digital preservation, the obsolescence of electronic formats.

- *licensing & patents*: FITS format has no restrictions on its use neither now nor in the future; there are no royalties (and never there will be) to pay for the use of the format, so anyone can use it and create software applications able to manage FITS files;
- *transparency*: one of the main features of the FITS format is the fact that metadata are stored in human and readable ASCII format, so anyone can easily examine, using a simple text editor, the header unit of the file to get information on the file itself; furthermore, the information contained in the Data Unit are stored in a very simple way, without any compression; due to its simplicity it is possible, and will be in the future, to read a file in FITS format using basic software tools;
- *self-documentation*: FITS file format is self-documenting; the keywords that appear in the Header Unit aim to exhaustively document the file and provide information such as size, origins, history data and anything else the creator wishes to enter; in addition to standard keywords it is also possible to use other keywords coined specifically to better describe the particular type of data that will be stored in the file;
- *self-containment*: all the information necessary to be able to correctly represent (on screen or print) a FITS file, are included in the file itself;
- *technical protection mechanisms*: the FITS format does not provide technical protection mechanisms; for example, encryption is not provided, nor are access control mechanisms such as usernames and/or passwords.

To better evaluate FITS format, it is possible to consider other requirements, such as the following:

- *robustness*: due to its simplicity and to the fact that it is not compressed, the corruption of certain sequences of bits does not produce as a rule the loss of information content of the file and the degradation of the image remains generally within acceptable levels;
- *stability*: FITS format is very stable: over more than forty years only four versions have been released;
- *backward and forward compatibility*: the format is compatible backward and forward. Any changes to the format that make it no longer a valid FITS file are not allowed;
- *device independence*: being a multi-platform format, a file in FITS format can be viewed, printed or otherwise reproduced in a reliable and consistent way regardless of platform hardware and software used;
- *accessibility*: it is possible to define a keyword to define the "alternative text" that can be read by the visualization software to describe the image, and this is very useful in cases where the image is enjoyed by persons with impaired vision;

–	*unmodifiability*: FITS can be modified using specially designed software (FITS editor). However, if necessary, its modifiability cannot be ensured by the use of technical measures such as using checksums or digital signatures, etc.;
–	*security*: in the current state of knowledge, FITS cannot contain viruses or other forms of malware;
–	*efficiency*: FITS is non-compressed, so the size of files encoded in this format are often quite high.

In summary, FITS has many features that make it a very attractive format for digital preservation purposes.

5.0 Further developments

Although FITS is currently the de facto standard for data storage and transmission in astronomy and other emerging fields, in recent years some scholars have begun to consider FITS outdated, based on the consideration that the choices made at the time of its writing, though common at the time, are now limiting the usefulness of the format itself. In their view, today's astronomy relies more than ever on a diverse set of data models with complex metadata (Thomas et al. 2015). As a result, in 2015 the Advanced Scientific Data Format (or ASDF) was released under the patronage of the Space Telescope Science Institute (Greenfield et al. 2015). Its aim is to be, just like FITS, a hybrid text and binary format, containing both human editable metadata for interchange. It is designed for extensibility and structurally complex, and raw binary data, fast to be loaded and used. The overall objectives of ASDF are very ambitious:[13]

–	it has a hierarchical metadata structure, made up of basic dynamic data types such as strings, numbers, lists and mappings;
–	it has human-readable metadata that can be edited directly in place in the file;
–	the structure of the data can be automatically validated using schema;
–	it is designed for extensibility: new conventions may be used without breaking backward compatibility with tools that do not understand those conventions. Versioning systems are used to prevent conflicting with alternative conventions;

13	Except from: "ASDF Standard. Introduction," last updated May 10, 2021, https://asdf-standard.readthedocs.io/en/1.6.0/intro.html. The latest version of ASDF standard (v. 1.6.0) is available at https://asdf-standard.readthedocs.io/en/1.6.0.

– the binary array data (when compression is not used) is a raw memory dump, and techniques such as memory mapping can be used to efficiently access it;
– it is possible to read and write the file in as a stream, without requiring random access;
– it is built on top of industry standards, such as YAML and JSON Schema to take advantage of a larger community working on the core problems of data representation. This also makes it easier to support ASDF in new programming languages and environments by building on top of existing libraries;
– since every ASDF file has the version of the specification to which it is written, it will be possible, through careful planning, to evolve the ASDF format over time, allowing for files that use new features while retaining backward compatibility with older tools.

The basic structure of ASDF files is straightforward, consisting of three sections:

– an *header* (mandatory) that indicates that the file of interest is in ASDF data format, stating clearly which version of the standard is used;
– a *tree* (optional), i.e. a YAML segment that provides a single structured view of all the data in the file. It is the main part of most ASDF files;
– zero or more *binary blocks* (optional) that are just a cluster of binary data, allowing for a simple and flexible type system.

According to its creators, ASDF is intended to go much beyond the cases previously handled by FITS by providing more descriptive keyword names, greater flexibility towards data values, an intrinsic hierarchical structure, and the possibility of sharing references to the same objects even between different elements. The flexibility guaranteed by the two data formats allows for a great deal of interchangeability: if a specific dataset must be produced in FITS, it is possible to embed ASDF in FITS, thanks to the use of the EXT-NAME = 'ASDF' extension. Vice versa, it is also possible to store a FITS file in ASDF, given the fact that the structure of the latter is more complex than the former.[14] Giving these premises, this fairly young digital data format seems to have what it takes to be a worthy heir of FITS, but only time can tell if this will be the case.

---

14 See Appendix A to ASDF Specification: "Appendix A. Embedding ASDF in FITS," last updated May 10, 2021, https://asdf-standard.readthedocs.io/en/1.6.0/asdf_in_fits.html.

6.0 Conclusions

FITS is a versatile and general-purpose file format and can archive and preserve many types of research data, including bitmaps, ASCII text, binary tables, multidimensional arrays such as data cube (3D) (e.g. temperature varying with time) or hypercube (4D, 5D, …) (e.g. temperature varying with time and altitude). It is extensible: user communities can develop specific conventions that refine the FITS standard in order to adapt it to specific contexts. Thanks to its versatility and extensibility, somebody proposed renaming FITS as "Flexible Image and Table Systems" (instead of "Flexible Image Transport Systems").

FITS meets a number of desirable requirements for a format suitable for digital preservation, and it is also one of the few formats that has stood the test of time because, more than forty years after its inception, it is still widely used. FITS was designed with an eye towards long-term archival preservation, and the maxim "Once FITS, always FITS" that has been coined for it clearly establishes that developments to the format must be backward compatible and that FITS files will never be made obsolete by more recent versions.

Furthermore, the Italian Unification Body (UNI, Ente Nazionale di Unificazione) is in the process of publishing the standard "Management Processes for the long-term preservation of digital images using the FITS format" (code name: UNI1606754). This new standard, which is likely to be released in 2022, will certainly contribute to its greater adoption. At ISO (International Standardization Organization) there is an ongoing proposal for its standardization as well.

Thus, FITS should be seriously considered as a format for long-term archiving and preservation of scientific data, at least in some areas of the EO field.

References

Albani, Mirko. 2012. "Long term preservation of earth observation data: The challenge and the cooperation activities". In *IEEE International Geoscience and Remote Sensing Symposium*, *(IGARSS 2012), 22-27 July 2012*, 7279-82.

Albani, Sergio, and David Giaretta. 2009. "Long-term Preservation of Earth Observation Data and Knowledge in ESA through CASPAR". *International Journal of Digital Curation* 4, no. 3: 4-16.

Allegrezza, Stefano. 2011. "Analisi del formato FITS per la conservazione a lungo termine dei manoscritti. Il caso significativo del progetto della Biblioteca Apostolica Vaticana", *DigItalia*, no. 2.

Ammenti, Luciano, 2012, "BAV and the FITS (Flexible Image Transport System) format 40 Years of Experience in Long-Term Digital conservation". In *European Week of Astronomy and Space Science (EWASS), 1-6 July 2012 Rome, Italy*. https://www. vatlib.it/moduli/Ammenti_EWASS2012.pdf.

"Appendix A. Embedding ASDF in FITS." Last updated May 10, 2021. https:// asdf-standard.readthedocs. io/en/1.6.0/asdf_in_fits.html.

"ASDF Standard. Introduction." Last updated May 10, 2021. https://asdf-standard. readthedocs.io/en/1.6.0/intro.html.

Conway, Esther, S. Pepler, Wendy Garland, D. Hooper, F. Marelli, L. Liberti, E. Piervitali, K. Molch, H. Glaves, and L. Badiali. 2014. "Ensuring the Long Term Impact of Earth Science Data through Data Curation and Preservation". *Information Standards Quarterly* 25, no. 3: 28. https://doi: 10.3789/isqv25no3.2013.05.

COSMO. "GRIB Data Format used for tge COSMO-Model System Features of GRIB." Last accessed December 13, 2021. http://www.cosmo-model.org/content/ model/documentation/grib/grib_ features.htm.

ESA (European Space Agency), and CEOS/WGISS. 2015. "Earth Observation Preserved Data Set Content (PDSC)", https://ceos.org/document_management/ Working_Groups/WGISS/Interest_Groups/Data_Stewardship/Recommenda tions/EO%20Preserved%20Data%20Set% 20Content_v1.0.pdf.

ESA (European Space Agency). 2015. "Long Term Preservation of Earth Observation Space Data. Preservation Guidelines". https://earth.esa.int/eogateway/docu ments/20142/37627/ EO-Data-Preservation-Guidelines.pdf.

ESA, "SAFE." Last accessed December 13, 2021. https://earth.esa.int/SAFE/index. html.

Greenfield, Perry E., Michael Droettboom, and Erik M. Bray. 2015. "ASDF: A new data format for astronomy". *Astronomy and Computing* 12: 240-51. https://www. sciencedirect.com/science/article/pii/S2213133715000645.

Greisen, Eric W., and Ronald H. Harten.1981. "An Extension of FITS for Groups of Small Arrays of Data". *Astronomy & Astrophysics Supplement Series* 44, 371.

Greisen, Eric W., Donald C. Wells, and Ronald H. Harten. 1980. "The FITS Tape Formats: Flexible Image Transport Systems". In *Applications of Digital Image Processing to Astronomy, Proceedings*, edited by D.A. Elliott. Society of Photo-Optical Instrumentation Engineers (SPIE) Conference Series 264, 298. https://doi. org/10.1117/12.959819.

Grosbol, Preben. 1988. "The FITS Data Format". *Bulletin d'Information du Centre de Donnees Stellaires*, no. 35: 7. https://adsabs.harvard.edu/full/1988BICDS..35....7G.

Hanisch, Robert J., A. Farris, E. W. Greisen, W. D. Pence, B. M. Schlesinger, P. J. Teuben, R. W. Thompson, and A. Warnock III. 2001. "Definition of the Flexible Image Transport System (FITS)". *Astronomy & Astrophysics* 376: 359–380. https:// doi:10.1051/0004-6361:20010923.

ISO (International Standardization Organization). 2020. "ISO 19165-2:2020 — Geographic information — Preservation of digital data and metadata — Part 2: Content specifications for Earth observation data and derived digital products". https://www.iso.org/standard/ 73810.html.

Library of Congress. 2021. "Sustainability of Digital Formats: Planning for Library of Congress Collections", https://www.loc.gov/preservation/digital/formats.

Manoni, Paola, Ángela Núñez Gaitán, and Irmgard Schuler. 2018. "The Vatican Library's Digital Preservation Project", Paper presented at: IFLA WLIC 2018 – Kuala Lumpur, Malaysia – Transform Libraries, Transform Societies in Session 160 – Preservation and Conservation with Information Technology. http://library.ifla.org/id/eprint/2113/1/160-manoni-en.pdf.

Molch, Katrin, Rosemarie Leone, Mirko Albani, and Eberhard Mikusch. 2012. "User needs and requirements impacting the long term preservation of earth observation data". In *IEEE International Geoscience and Remote Sensing Symposium, (IGARSS 2012)*: 7283-85, https://doi: 10.1109/IGARSS.2012.6351980.

NASA. "FITS Standard Document." Last accessed December 13, 2021. https://fits.gsfc.nasa.gov/fits_standard.html.

NASA. "Space Physics Data Facility." Last accessed December 13, 2021. https://cdf.gsfc.nasa.gov/html/FAQ.html.

NASA. 2011. "NASA Earth Science Preservation Content Specification". Greenbelt, Maryland: Goddard Space Flight Center, https://cdn.earthdata.nasa.gov/conduit/upload/10607/NASA_ESD_Preservation_Spec.pdf.

NASA. 2014. "A Primer on the FITS Data Format." Last revised October 28, 2014. https://fits.gsfc.nasa.gov/fits_primer.html.

NASA/GSFC Astrophysics Data Facility. 1997. "A User's Guide for the Flexible Image Transport System (FITS). Version 4.0, 1997", April 14, 1997. https://fits.gsfc.nasa.gov/users_guide/usersguide.pdf.

National Research Council. 1995. "Preserving Scientific Data on Our Physical Universe: A New Strategy for Archiving the Nation's Scientific Information Resources". Washington, DC: The National Academies Press, 11. https://doi.org/10.17226/4871.

NCEI. "Global Forecast System." Last accessed December 13, 2021. https://www.ncei.noaa.gov/products/weather-climate-models/global-forecast.

Pinna, Gian Maria, and Stéphane Mbaye. 2014. "SAFE Archive for ENVISAT and Third Party Mission Products". July 17, 2014. https://earth.esa.int/workshops/envisatsymposium/authors/CXNL_07A03_507957.html.

Ramapriyan, Hampapuram, Dawn Lowe, and Kevin Murphy. 2017. "Towards a preservation content standard for earth observation data". In *Proceedings of the 2017 conference on Big Data from Space (BiDS' 2017) 28th-30th November 2017 Toulouse (France)*. https://ntrs.nasa.gov/citations/20170011347.

STSI. "ASDF Standard." Last accessed December 13, 2021. https://asdf-standard.readthedocs.io/en/1.6.0.

Thomas, Brian, Tim Jenness, Frossie Economou, Perry Greenfield, Paul Hirst, David S. Berry, Erik Bray, Norman Gray, Demitri Muna, and James Turner. 2015. "Learning from FITS: Limitations in use in modern astronomical research", *Astronomy and Computing* 12, 133-45. https://arxiv.org/abs/1502.00996.

Unidata. "Network Common data Form (NetCDF)." Last accessed December 13, 2021. https://www.unidata.ucar.edu/software/netcdf.

Wells, Donald C., and Eric W. Greisen. 1979. "FITS: a Flexible Image Transport System". In *Image Processing in Astronomy*, edited by G. Sedmak, M. Capaccioli, R. J. Allen, 445.

Wells, Donald C., Eric W. Greisen, and Ronald H. Harten. 1981. "FITS: a Flexible Image Transport System", *Astronomy & Astrophysics Supplement Series*, no. 44. https://articles.adsabs.harvard.edu/pdf/1981A%26AS...44..363W.

An ontology for the representation of Earth Observation data: a step towards semantic interoperability[1]

Giovanna Aracri
IIT-CNR, Italy

Assunta Caruso
University of Calabria, Italy

Antonietta Folino
University of Calabria, Italy

Abstract

Earth observation (EO) consists in collecting information related to the Earth's physical, chemical and biological systems. Data are gathered through remote sensing technologies, mainly consisting of satellites, and represent an essential source of information, which can be associated with administrative, social and economic issues in order to support policy analysis and decision-making. In recent years, characterized by an intensification of climate change phenomena and of extreme environmental events, the amount of EO data collected has significantly increased. Nevertheless, their huge volume and heterogeneity do not allow the data to be easily and promptly used by the scientific community. The different methods adopted for collecting, processing, cataloguing and describing data through metadata introduce an additional high level of variability. In this sense, it is important to guarantee technical, syntactic and semantic interoperability among data. This paper focuses on semantic interoperability issues in the EO domain and introduces an ontological representation of knowledge tailored to the concept of Essential Variables (EVs). The ontological model has been defined within a European research program which is also oriented towards the development of a Knowledge Base in the specific domain. Its general intent, however, is to provide a framework concerning a set of EVs, identified and characterized by a community of experts in order to guarantee information and knowledge generation from observable environmental data.

---

1   Authors have equally contributed to this work, however Giovanna Aracri particularly focused on "The method", "Definition of the ontology" and "The use case description"; Assunta Caruso focused on "Introduction" and "Conclusion"; Antonietta Folino focused on "Literature Review" and "Interoperability with other vocabularies".

1.0 Introduction

The concept of interoperability is increasingly at the heart of several initiatives promoted by the European Union. The main aim is to improve interaction among Member States and effective communication among digital devices, networks and data repositories (Directorate-General for Informatics of the European Commission 2017). The European Interoperability Framework (EIF) defines the concept of interoperability as follows:

> "the ability of organizations to interact towards mutually beneficial goals, involving the sharing of information and knowledge [...] by means of the exchange of data between their ICT systems" (European Commission 2010, 2).

identifies four levels of interoperability: legal, organizational, semantic and technical. A significant implementation of this general framework is represented by the INSPIRE Directive (2007), aimed at establishing:

> "the Infrastructure for Spatial Information in the European Community [...], for the purposes of Community environmental policies and policies or activities which may have an impact on the environment." (European Parliament and Council 2007, 4).

Policy and decision making in this specific domain cannot ignore the harmonization of data which is produced and exchanged, nor the accomplishment of interoperability at all levels. This paper deals with semantic interoperability issues and encourages the use of standards and specifications to preserve the precise meaning of information during communication. The fulfilment of this objective is based on both syntactic and semantic aspects and requires the development of vocabularies and data models to describe and represent data. Many efforts have already been made to improve syntactic interoperability, such as standardization of data formats and development of XML-based data encoding rules, i.e. an ISO (International Organization for Standardization) standard and an OGC (Open Geospatial Consortium) standard (Nagai et al. 2012). However, the true challenge is that of achieving a significant level of semantic interoperability, intended as "the ability of different agents, services, and applications to communicate [...] data, information, and knowledge – while ensuring accuracy and preserving the meaning of that same data, information, and knowledge" (Zeng and Chan 2015). As affirmed by Zeng (2019) "with semantic interoperability, the expanded notion of data includes semantics and context, thereby transforming data into information".

The well-known data-information-knowledge-wisdom (DIKW) pyramid depicts a clear, complete and practical view concerning the transformation from data to wisdom (Frické 2018). Figure 1 offers a clear representation of

Figure 1. DIKW pyramid

how to handle the transformation from data to wisdom and shows the recognizable artefacts: (a) Data: Earth observations and measurements; (b) Information products: Essential Variables (EVs) and Indicators; (c) Knowledge products: indexes; (d) Wisdom actions: Sustainable Development Goals (SDGs).

This framework represents how raw data – thanks to further enhancement – evolve, support the accomplishment of the desired outcomes and allow impact assessment. Making the boundaries and the interlinking amongst these four layers explicit, helps us to encompass the unclear distinction between them and their meanings. This confusion is due to several definitions that have been provided by the scientific community, as well as different ways of interrelating their meaning. An attempt to clarify them has been made by Liew (2007) in a comprehensive dissertation based on the comparison of some definitions that have been used over time. In the specific field of Information Science and Knowledge Management and Engineering the explicit distinction of these key concepts is essential in order to avoid meaning overlap. Therefore, starting from the widely accepted assumption that in order for data to become useful and exploitable and be turned into usable information and knowledge, data need to be interpreted and enriched. The community of information scientists have developed different techniques and methodologies to carry out this transformation process (Zins 2007). Indeed, information is an added-value product generated by understanding data and working out relations among them and with physical and/or social phenomena (Craglia and Nativi 2018). Understanding information and working out valuable patterns generates knowledge in turn.

In this paper we will describe our contribution towards semantic interoperability through the definition of an ontological model useful to solve semantic mismatch of data.[2] The aim is to support the Virtual Earth Laborato-

---

2   The study described in this paper has been conducted within the ERA-PLANET Program – The European network for observing our changing planet – Call: H2020-

ry (VLab) in workflow execution (Nativi et al. 2019) by focusing on Essential Variables (EVs) linked to selected Societal Benefit Areas (SBAs),[3] improve the sustainability of EO-based indicator systems and inform the Sustainable Development Goals (SDGs). Several EVs (e.g. climate, water, energy, food, and biodiversity) have been defined to describe and represent knowledge and make it machine accessible and to leverage heterogeneity of data deriving from diverse sources (Buttigieg et al. 2019).

2.0 Literature Review

This section introduces some of the existing ontologies covering the environment domain by focusing on the themes they deal with and on their functions and real applications. The aim is to identify and to evaluate potential similar resources that could be reused and/or matched with the ontology under construction in the domain of EVs. In this perspective, a significant ontology is represented by the Sustainable Development Goals Interface Ontology (SDGIO), under development by UNEP (United Nations Environment Program) in collaboration with experts in the domain of knowledge representation.[4] The objective of the SDGIO is to logically represent and define entities relevant for the Sustainable Development Goals (SDGs) so that their meaning can be unambiguously understood and interpreted by the community of experts. Its importance here lies in the fact that it is tied both to the domain of interest and to the similarity of the aims of the ontology we are developing. Some concept definitions are not universally accepted or differ based on the context. Consequently, this can compromise the quality of data and the correct measurement of progress towards the corresponding targets. To this end, concepts included in the ontology will be mapped to the corresponding terminology in resources such as the UN System Data Catalogue and the SDG Innovation Platform. The SDGIO "aims to provide a semantic bridge between 1) the Sustainable Development Goals, their targets, and indicators and 2) the large array of entities they refer to".[5] Furthermore, the objective of the SDGIO is to provide, when available, dif-

---

SC5-2015-one-stage; Topic: SC5-15-2015; Type of action: ERA-NET-Cofund; Grant Agreement n. 689443. More specifically, this paper is focused on the GEOEssential Variables workflows for resource efficiency and environmental management project.

3   Group on Earth Observations (GEO), "Geo at a Glance," last accessed October 1, 2021, https://earthobservations.org/geo_wwd.php.
4   UNEP was requested to develop SDGIO by the IAEG-SDG (Inter-agency and Expert Group on SDG Indicators) during its 2nd meeting held in Bangkok in October 2015.
5   Ontobee, "Sustainable Development Goals Interface Ontology," last accessed October 1, 2021, http://www.ontobee.org/ontology/SDGIO.

ferent definitions for each concept, rather than a unique definition that would require member states to change their understanding of term meanings. This should guarantee coherence and prevent confusion in data handling, policy decision making and information management. Apart from the use of the SDGIO in local data systems and projects (i.e. in India, Germany and Japan),[6] it has been implemented on the UNEPLive portal (http:// uneplive.unep.org/) and it represents a useful support for UN statisticians and researchers as a reporting and monitoring solution. A more precise and consistent representation of knowledge about SDGs will help in "monitoring the status of how various targets and goals are being addressed around the world".[7] Currently the SDGIO is structured as follows: 514 classes, 144 object properties, 27 annotation properties and 702 instances.[8] The SDGIO is continuously updated, hence, new classes both strictly representative of SDGs as well as those concerning other related domains, are imported from other existing ontologies (e.g. ENVO for environment and climate, CHEBI for chemicals and waste, OBI for measurement, data collection and monitoring, PCO for populations and communities)[9] and are mapped to the concepts contained in GEMET (General Multilingual Environmental Thesaurus, https://www.eionet.europa.eu/gemet/en/themes/) in order to provide a more comprehensive and precise representation of the domain and to guarantee greater interoperability. Some other domains not yet covered by existing ontologies (i.e. human rights or financial measures), as well as some regional understandings, would need better coverage, therefore the SDGIO will be further developed to include new knowledge. In this sense, a list of candidate terms needing a definition already exists. The SDGIO and the ontology discussed in this paper undoubtedly share some elements: for both, SDGs and their targets and indicators represent relevant concepts; both aim at supporting local and global policy and decision makers in adopting strategies to monitor the human impact on the environment by providing them with relevant and consistent knowledge (Buttigieg et al. 2016a). Therefore, it is worthwhile to take the SGDIO into consideration when modelling the EV

---

6   Coppens, Ludgarde and Dany Ghafari, "Sustainable Development Goals Interface Ontology (SDGIO). Progress," last accessed October 1, 2021, https://unstats.un.org/ unsd/unsystem/Documents-Sept2018/Presentation-SDGIdentifiers-UNEP.pdf.

7   Jennifer Zaino, "Ontology Plays a Part in United Nations Sustainable Development Goals Project," last accessed October 1, 2021, https://www.dataversity.net/ontology-has-big-part-to-play-in-united-nations-sustainable-development-goals-project/#.

8   SDGIO is registered both in OntoBee (http://www.ontobee.org/ontology/SDGIO), a linked data server for ontologies, and in EBI OLS, an ontology lookup service (https://www.ebi.ac.uk/ols/ontologies/sdgio).

9   GitHub, "Domain ontologies relevant to SDGIO," last accessed October 1, 2021, https://github.com/SDG-InterfaceOntology/sdgio/wiki/Domain-ontologies-relevant-to-SDGIO.

ontology and in formalizing explicit links between them through ontology mapping techniques (Ding and Foo 2002). This will allow the two systems to be kept independent without interfering with each other's purposes. Nevertheless, the specific aims of the GEOEssential project, require that our ontology includes a specific focus on concepts such as EVs, Observables, Datasets, and so on, in order to fulfill the specific requests of the communities of experts and to be integrated within the VLab in order to run workflows. In this regard, the ongoing ontology is more project-oriented, while the SDG-IO conceptual model is based on more general classes, represented by a core set of universal terms (e.g. process, role, entity, etc.). As mentioned above, the SDGIO project is based on the reuse of existing ontologies. In particular, it adopts and imports the ENVO conceptual model addressing it towards the evaluation of the sustainability of human actions (Buttigieg et al. 2016b). ENVO (http://www.environmentontology.org) is a community-led, open project which seeks to provide an ontology for specifying a wide range of environments relevant to multiple life science disciplines and, through an open participation model, to accommodate the terminological requirements of all those needing to annotate data and to search datasets using ontology classes. ENVO is comprised of classes referring to key environment-types that may be used to facilitate the retrieval and integration of a broad range of biological data. At the moment of its development, in 2013, it represented concepts mainly belonging to biomes, environmental features, environmental materials; more recently it has grown in order to represent multiple fields related to the environment (e.g. habitats, environmental processes, anthropogenic environments, environmental health initiatives, concepts concerning the global Sustainable Development Agenda for 2030) (Buttigieg et al. 2016b). In constructing ENVO, the developers recognized the many existing resources which address, among other entities, environment-types and were motivated by the value of unifying such resources in a foundational – or building block – ontology developed within a federated framework and exclusively concerned with the specification of environment types, independent of any particular application. Classes describing natural environments currently dominate ENVO's content as the ontology is geared towards use in the biological domain. Nevertheless, ENVO is suitable for the annotation and search of any record that has an environmental component. ENVO is interoperable with the existing ontologies in the Open Biological and Biomedical Ontologies (OBO) Foundry and Library (http://www.obo-foundry.org/). Another ontology worth mentioning is SWEET (Semantic Web for Earth and Environment Technology Ontology)[10] which was origi-

---

10 BioPortal, "Semantic Web for Earth and Environment Technology Ontology," last accessed October 1, 2021, https://bioportal.bioontology.org/ontologies/SWEET.

nally developed by NASA Jet Propulsion Labs and is now under the governance of the ESIP (Earth Science Information Partners) Foundation. It is an example of a highly modular ontology suite which includes about 7,000 elements (Classes, Properties, Individuals, etc.) in 200 separate ontologies[11] covering Earth system science. A modular ontology is defined as a set of ontology modules, where these modules can be integrated through various proposed formalisms (Ensan et al. 2010). Indeed, SWEET is a mid-level ontology and consists of nine top-level concepts[12] that can be used as a foundation for domain-specific ontologies that extend these top-level SWEET components. SWEET has its own domain-specific ontologies, which extend the mid-level ontologies. The former can provide users interested in developing a finer-grained ontological framework for a particular domain with an initial solid set of concepts. Other relevant ontologies related to the environment domain and published in the form of Linked Open Data (LOD) can be found on the Linked Open Vocabularies (LOV) portal, an observatory of the semantic vocabularies ecosystem.[13] Some of the covered domains are: climate and forecasting, paleoclimatology, energy efficiency, living species, smart cities and homes, sensors and observations, emissions.

3.0 The method

The urgent need to improve accuracy and quality of data coming from Earth Observation (EO) monitoring is due to several threats: the large volume and variety of the acquired dataset; the complexity with which data are expressed; the difficulty to understand which data need to be extracted and the different perspectives and conceptualizations adopted to develop the dataset and model framework. In this section, we will outline the main features of the ontological model, that is, the rules and the constraints that have been followed to develop it. As already mentioned, the aim is to provide a representative description of the EO domain with a specific focus on Essential Variables (EVs) necessary for the GEO infrastructure to derive policy relevant indicators in order to contribute to the continuous improvement and application of interoperability within it. This is quite challenging because EO monitoring systems produce a myriad of valuable, complex and

---

11   Numbers are accurate as of October 2021.

12   e.g. Representation (math, space, time, data); Realm (Ocean, Land, Surface, Terrestrial Hydrosphere, Atmosphere, etc.); Phenomena (macro-scale ecological and physical); Processes (micro-scale physical, biological, chemical, and mathematical), Human Activities (Decision, Commerce, Jurisdiction, Environmental, Research).

13   Linked Open Vocabularies (LOV), "VOCABS all you know about lov!," last accessed October 1, 2021, https://lov.linkeddata.es/dataset/lov/vocabs?&tag=Environment.

heterogeneous data and information that need to be managed, controlled and interpreted to become helpful for policy makers to solve problems more effectively. The ontology design is a complex modelling task and reasonably requires the continuous interaction of both experts in the specific domain and experts in the use of domain-specific KOSs. This collaboration is essential in defining the abstraction layer which is useful to implement in an information system. Ontology engineering consists in several steps: information collection, identification of the relevant concepts to include in the ontology, definition of classes, sub-classes and class instances, description of the semantic relationships by means of properties and axioms through rules, constraints and restrictions (Chantrapornchai and Choksuchat 2016).

3.1 Definition of the ontology

As is well known, an ontology is a Knowledge Representation System, considered as the main technology of the Semantic Web and of several other applicative contexts (e.g. e-commerce, problem solving, data integration, etc.). It provides a formalized and accepted conceptualization of a domain (Gruber 1995) guaranteeing common information sharing and understanding, reuse of the modelled knowledge and advanced capability of reasoning and making assumptions (Gomez-Perez and Benjamins 1999). In modelling an ontology, the main concepts of the domain are represented through classes, further subdivided into sub-classes according to hierarchical relationships. The taxonomy granularity depends on the information gathered and on the kind of data that needs to be aggregated. In order to achieve a more expressive representation, especially if compared with other KOSs (Kister et al. 2011), alongside these hierarchical arrangements other types of relationships between classes can be provided and explicitly expressed by means of binary typed object properties. In our ontology, the most generic level of conceptualization is represented by both: classes which support and start the EV generation process (e.g. Algorithm, Dataset, Method of computation, etc.) and classes that are EO-centred (e.g. Essential Variable, Policy Goal, Indicator, etc.) and which therefore provide a representative – albeit not exhaustive – overview of this specific domain. In turn, classes are organized into subclasses according to hierarchical principles thereby introducing superordinate and subordinate levels. Two kinds of hierarchical relations are expressed in the ontology: the generic relationship (also known as Is-a) and the partitive relationship (also known as whole part or type-of). The former specifies a connection between a class and its members and fulfills the *all-and-some test* (ISO 25964-1:2011, 59) (e.g. Observable → Land cover); the latter, on the other hand, states that the superordinate concept (whole) includes one or more subordinate concepts (parts) (e.g. Urban area →

City). As already explained, Essential Variable is a crucial class in the model because it represents the main output in the EV generation process. Therefore, it is necessary to detail all the relevant information useful for describing it. The EV class is organized at different hierarchical levels and provides a significant representation of all the main aspects regarding this concept (e.g. Essential Variable → Essential climate variable → Land cover). Knowledge concerning EVs is extremely dynamic and needs to be continuously monitored, because many EVs have been identified by a panel of experts and are an established reality (such as Essential climate variables, Essential Ocean Variables, etc.),[14] others, on the other hand, have not yet been consolidated and shared by the community of experts. This debate does not influence the intention of the EV class, which is however inclusive and ready to welcome further Evs.[15]

Figure 2. Essential Variable taxonomy

---

14   Earth Data Open Access for Open Science, "Essential Variables," last accessed October 1, 2021, https://earthdata.nasa.gov/learn/backgrounders/essential-variables.

15   Some other EVs will be included in the taxonomy, following the ongoing research conducted by the project partners (ex. Essential Land Variables, cross domain EVs, etc.).

Besides the hierarchical relations based on the inheritance principle, according to which sub-classes must satisfy all characteristics of the class immediately above it, logical connections can be explicitly expressed by means of the so-called object properties while some attributes which provide additional details are expressed through data properties.

The formalization of these biunivocal relationships (direct and inverse), make several statements explicit in the form of subject-predicate-object triple (e.g. "Indicator 15.3.1 measures Target 15.1" and vice versa "Target 15.1 isMeasuredBy Indicator 15.3.1"). A single statement interlinked with other statements, creates a rich and interconnected structure which unambiguously represents the conceptualization of the domain and provides the narrative of the ontology with regard to the specific it addresses.

Figure 3. Relationships between classes

Object and Data Properties broaden the explicitation of the technical semantic associations among concepts with respect to the information contained in the domain-documentation set. The general ontology model has been tailored according to use cases regarding some SDG indicators. Given the formalism that characterizes the ontologies' configuration in the way these tools represent a determined specialized field of knowledge, the typologies of these semantic correlations had to be set according to a high abstract perspective. Indeed, in the development of the main connections existing among the domain-oriented concepts one should take into account the openness trait in the *domain* and *range* declarations, meaning that it is advisable to keep the structure as simple as possible to avoid errors in the inference process and to use conjunction structures to let concepts share some properties, (e.g., *Land_Surface_Temperature UsedToDetect Peat_fire **or** Wildfire*). In fact, inference processes can benefit from the formal declaration of restrictions and axioms that clearly define which and how specific individuals can be related with each other. The analysis of the documentation set that has constituted the starting point from which to populate the main domain-targeted concepts and their relationships in the ontological taxonomy represented by the object and data properties in OWL language (Petasis et al. 2011). In this specific case, the application falls within a highly specialized domain of study that is made up of several fragmented documents that provide essential information to be matched in order to provide a picture of the technical documentation. The definition of a taxonomy as an ontology supports the achievement of this systematization.

3.2 The Use case description

In order to assess the level of semantic interoperability achieved thanks to the ontology and concurrently validate its consistency, several use-cases have been integrated and acquired in the VLab. To verify if the specific issues regarding the EV domain have been considered and are well represented in the model, and also to test the consistency of the ontology compared to the GEOEssential objectives, the SDG Indicator 15.3.1 "Proportion of land that is degraded over total land area"[23] has been pointed out as a use-case. This Indicator has been recognized as relevant for the simulation and adopted as a use case both because it has been used in other experimentations within the project and because it allows to investigate a real and currently interesting phenomenon (i.e. Land degradation) that occurs in the domain. Indeed, concerning the first aspect, a workflow related to the specific topic of Land Degradation has been modelled by some project partners and was successfully tested for running within the VLab[16] (Giuliani et al. 2020). Consequently, it has been necessary to include specific information in our ontological model. The importance of this issue for the scientific community depends on the fact that this process is "undermining the well-being of 3.2 billion people, driving species extinction, intensifying climate change, leading to increased risk of migration and conflict" and that 75% of the

---

16   Information are also available on the GEOEssential Dashboard: https://geoessential. unepgrid.ch/mapstore/#/dashboard/36. The workflow modeled by the project partners contains the following interesting information: EVs uses (Land cover, land productivity, soil carbon); Inputs (Landsat, Modis, Copernicus, ESA-CCI-LC, HWSD, SolidGrid250m, Global SOC Map); Outputs (Land degradation indicators); Main Process (Trends.Earth model: http://trends.earth).

Earth's land areas are substantially degraded and the percentage will reach 90% by 2050.[17] In order to improve and specialize the ontology structure so that it could represent specific subjects related to Land degradation, various authoritative documents, mainly taken from the United Nations website, were consulted and analyzed. Other valuable information has been provided by a panel of experts involved in the project with the abovementioned partners. Nevertheless, the accuracy and the completeness of the model are not fully guaranteed at the moment, as further validation is being carried out by domain experts and other potential suggestions will come from the running of workflows within the VLab. In fact, the involvement of experts, both from a technical and from a domain point of view, is mandatory for the development of such a system, especially since all the information modelled will be used by the decision and policy makers to select suitable actions which will allow to reach the objectives expected by the specific SDGs.

The knowledge regarding Indicator 15.3.1 currently formalized and available in the ontological model can be summarized as follows:

- the corresponding Target has been specified (Indicator 15.3.1 measures Target 15.3), as well as the Goal referred to the Target (Goal 15 hasTarget Target 15.3);
- the related sub-indicators have been listed;
- the datasets providing the data useful for the computation of the Indicator have been listed and linked to the model they are able to generate (e.g. GIMMS generates MOD13Q1, which is an EV generation model);
- Essential Variables potentially related to the Indicator have been identified and organized according to the corresponding Category (e.g. Precipitation is an Essential Climate Variable and belongs to the Atmosphere category);
- the Indicator has been related to the method of computation generally used to calculate it (One out, all out);
- the Model class specifies both EV generation models and Indicator generation models (Trends.Earth).

Indicator 15.3.1 has been linked to various other indicators, some of which belong to other SDGs (e.g. Indicator 15.3.1 isRelatedTo Indicator 6.6.1 "Change in the extent of water-related ecosystems over time").

Using explanatory case studies allows us to further test the ontological model or to improve it if some important information is missing or is not correctly modelled.

---

17   European Commission, "Land degradation threatens the well-being of people and the planet," last accessed October 1, 2021, https://ec.europa.eu/jrc/en/science-update/land-degradation-threatens-well-being-people-and-planet.

Figure 4. Indicator 15.3.1 relationships

3.3 Interoperability with other vocabularies

The ontology aims at describing shared knowledge in the EO community by exploiting the potential of using a knowledge representation system able to ensure a suitable level of formalization and explication. In building and enhancing the ontology other existing and reliable references – in particular those described in section 2.0 have also been taken into consideration. Indeed, reusing concepts reinforces semantic interoperability and discoverability also in line with FAIR data principles. This point of view has been largely applied in both information identification and collection tips to select the most relevant concepts in the domain, and in cross-vocabulary alignment. In particular, the ontology schema has been enriched by the inclusion – in the form of classes and subclasses – of concepts represented by terms obtained in a preliminary phase of term extraction from existing vocabularies and from a specific corpus (Aracri et al. 2020). Furthermore, the ontology construction process is based on an incremental method, consisting in the gradual enrichment of the general model through the analysis and the consequent representation of specific use cases, which allow to test the goodness of the model itself.

As for the alignment with external vocabularies, several explicit mappings have been established by means of the OWL SKOS Model[18] and in particular by the Annotation properties 'exactMatch', 'closeMatch', 'broaderMatch', 'narrowerMatch', 'relatedMatch'. As an example, Figure 5 shows the close equivalence between Water Vapour in our ontology and Water Vapour in ENVO. The choice of the degree of correspondence depends on the different taxonomies in which these concepts are included: in the first case it is a subclass of Essential Climate Variable, thus it is intended with this specific meaning, reflected in the provided definition, in the second case it is a subclass of Gaseous environmental material, therefore it has a more comprehensive meaning.

Figure 5. Example of Close Match

Vocabulary alignment is a key process towards semantic interoperability because it allows to maintain the independence of the vocabularies involved for two main reasons: changes in one of should not affect the others and each vocabulary can be autonomously used in its specific context and with its specific purpose because of its own conceptual structure.

4.0 Conclusion

Ensuring a high standard degree of technical and semantic interoperability in managing large amounts of data and turning them into shareable information and knowledge is an interesting challenge, which contemplates the involvement of interdisciplinary competences and expertise. This approach permits to achieve greater integration and interaction, ensures better results, that incorporate different perspectives and, as stated in (Kleineberg 2016) it generates "a superior understanding of a particular question or object of interest" and it allows to investigate "problems whose solutions are beyond the scope of a single domain". The present paper considers the concept of

---

18 World Wide Web Consortium (W3C), "SKOS Simple Knowledge Organization System – Reference," last accessed October 1, 2021, https://www.w3.org/TR/skos-reference/.

interoperability beyond the purely technical issue, and it is enhanced by the development of an ontological model, which is embedded in the Key Enabling Technologies (KETs). The open and interoperable access to data and knowledge is assured by the development of a comprehensive Knowledge Platform, and by tailoring general functionalities to the specific requirements, which in this case concerns the EV domain. The integration is experimented utilizing the presented ontological model within a platform in order to support the running of data, which are complex, heterogeneous and dispersed in several datasets. The implementation of these patterns and technologies will ensure full horizontal interoperability with relevant EO initiatives and programs (e.g. GEOSS, Copernicus). The main advantages deriving from the implementation of semantic services and from the organization of the knowledge domain in an ontological model are related to the improvement of the information retrieval process, both for experts and data providers as well as for policy makers, who should be able to take decisions and to adopt knowledge-based policies (Kornyshova and Deneckère 2012), and to the provision of advanced discovery and modelling services for answering complex queries.

Acknowledgments

This activity was funded by the European Commission in the framework of the program "The European network for observing our changing planet (ERA-PLANET)", Grant Agreement: 689443.

References

Aracri, Giovanna, Assunta Caruso, and Antonietta Folino. 2020. "An Ontological Model for Semantic Interoperability Within an Earth Observation Knowledge Base." In *Knowledge Organization at the Interface, Proceedings of the Sixteenth International ISKO Conference, Aalborg, Denmark, 6-8 July 2020*, edited by International Society for Knowledge Organization, Marianne Lykke, Tanja Svarre, Mette Skov and Daniel Martínez-Ávila, 17: 13-22. Aalborg, Denmark: Ergon. https://doi.org/10.5771/9783956507762-13.

BioPortal. "Semantic Web for Earth and Environment Technology Ontology." Last accessed October 1, 2021. https://bioportal.bioontology.org/ontologies/SWEET.

Buttigieg, Pier Luigi, Ramona L. Walls, and Anne Thessen. 2019. "Semantic Interoperability Solutions for the Essential Variables: Focus on biodiversity." In *BiodiversityNEXT, Leiden, The Netherlands, 22-25 October 2019*. https://doi.org/10.3897/biss.3.36234.

Buttigieg, Pier Luigi, Mark Jensen, Ramona L. Walls, and Cristopher J. Mungall. 2016a. "Environmental semantics for sustainable development in an interconnected biosphere." In *Proceedings of the International Conference on Biological Ontology and BioCreative (ICBO/BioCreative) Corvalis, Oregon, USA, 1-4 August 2016,* edited by Pankaj Jaiswal, Robert Hoehndorf, C. N. Arighi, A. Meier.

Buttigieg, Pier Luigi, Evangelos Pafilis, Suzanna E. Lewis, Mark P. Schildhauer, Ramona L. Walls, and Cristopher J. Mungall. 2016b. "The environment ontology in 2016: bridging domains with increased scope, semantic density, and interoperation." *Journal of Biomedical Semantics* 7, no. 57. https://doi.org/10.1186/s13326-016-0097-6.

Chantrapornchai, Chantana and Chidchanok Choksuchat. 2016. "Ontology construction and application in practice case study of health tourism in Thailand." *SpringerPlus* 5, no. 1: 1-31. https://doi.org/10.1186/s40064-016-3747-3.

Coppens, Ludgarde and Dany Ghafari. "Sustainable Development Goals Interface Ontology (SDGIO). Progress." Last accessed October 1, 2021. https://unstats.un.org/unsd/unsystem/Documents-Sept2018/Presentation-SDGIdentifiers-UNEP.pdf.

Craglia, Max and Stefano Nativi. 2018. "Mind the Gap: Big Data vs. Interoperability and Reproducibility of Science." *Earth Observation Open Science and Innovation,* edited by Pierre-Philippe Mathieu and Christoph Aubrecht. ISSI Scientific Report Series 15: 121-41. Springer Open. https://link.springer.com/content/pdf/10.1007%2F978-3-319-65633-5.pdf.

Ding, Ying and Shubert Foo. 2002. "Ontology research and development: Part 2 – a review of ontology mapping and evolving." *Journal of Information Science* 28: 375-88.

Directorate-General for Informatics of the European Commission. 2017. "New European interoperability framework: promoting seamless services and data flows for European public administrations, Publications Office." Luxembourg: Publication Office of the European Union. http://doi:10.2799/78681.

European Commission. "Land degradation threatens the well-being of people and the planet." Last accessed October 1, 2021. https://ec.europa.eu/jrc/en/science-update/land-degradation-threatens-well-being-people-and-planet.

Earth Data Open Acces for Open Science. "Essential Variables." Last accessed October 1, 2021. https://earthdata.nasa.gov/learn/backgrounders/essential-variables.

European Parliament and European Council. 2007. "Directive 2007/2/EC of the European Parliament and of the Council of 14 March 2007 establishing an Infrastructure for Spatial Information in the European Community (INSPIRE)".

Frické, Martin H. 2018. "Data-Information-Knowledge-Wisdom (DIKW) Pyramid, Framework, Continuum." In *Encyclopedia of Big Data*, edited by Laurie A. Schintler, Connie McNeely. Springer. https://doi.org/10.1007/978-3-319-32001-4.

GitHub. "Domain ontologies relevant to SDGIO." Last accessed October 1, 2021. https://github.com/SDG-InterfaceOntology/sdgio/wiki/Domain-ontologies-relevant-to-SDGIO.

Giuliani, Gregory, Paolo Mazzetti, Mattia Santoro, Stefano Nativi, Joost Van Bemmelen, Guido Colangeli, and Anthony Lehmann. 2020. "Knowledge generation using satellite earth observations to support sustainable development goals (SDG): A use case on Land degradation." *International Journal of Applied Earth Observation and Geoinformation* 88. https://doi.org/10.1016/j.jag.2020.102068.

Gomez-Perez, Asuncion and V. Richard Benjamins. 1999. "Overview of Knowledge Sharing and Reuse Components: Ontologies and Problem-Solving Methods." In *Proceedings of the IJCAI-99 workshop on Ontologies and Problem-Solving Methods (KRR5), 2 August 1999 Stockholm, Sweden,* edited by Asuncion Gomez-Perez, M. Gruninger, H. Stuckenschmidt, M.Uschold, 18.

Group on Earth Observations (GEO). "Geo at a Glance." Last accessed October 1, 2021. https://earthobservations.org/geo_wwd.php.

Gruber, Thomas R. 1995. "Toward Principles for the Design of Ontologies Used for Knowledge Sharing." *International Journal Human-Computer Studies* 43, 5-6: 907-28, https://doi.org/10.1006/ijhc.1995.1081.

ISO 25964-1:2011 Information and documentation — Thesauri and interoperability with other vocabularies — Part 1: Thesauri for information retrieval.

Kister, Laurence, Evelyne Jacquey and Bertrand Gaiffe. 2011. "Du thesaurus onto-ter-minologie: relations sémantiques vs relations ontologiques." *Corela (Cognition, Répresentation, Langage)* 9, no. 1. https://doi.org/10.4000/corela.1962.

Kleineberg, Michael. 2016. "Book Review: Interdisciplinary Knowledge Organization by Rick Szostak, Claudio Gnoli, and María López-Huertas." *Knowledge Organization* 43: 663-67.

Kornyshova, Elena and Rébecca Deneckère. 2012. "Using an Ontology for Modeling Decision-Making Knowledge." *Advances in Knowledge-Based and Intelligent Information and Engineering Systems, Frontiers in Artificial Intelligence and Applications,* 243: 1553 – 62. https://doi:10.3233/978-1-61499-105-2-1553.

Liew, Anthony. 2017. "Understanding Data, Information, Knowledge and their Inter-Relationships." *Journal of Knowledge Management Practice* 7, no. 2.

Linked Open Vocabularies (LOV). "VOCABS all you know about lov!" Last accessed October 1, 2021. https://lov.linkeddata.es/dataset/lov/vocabs?&tag=Environment.

Nagai, Masahiko, Masafumi Ono, and

Nativi, Stefano, Mattia Santoro, Gregory Giuliani, and Paolo Mazzetti. 2020. "Towards a knowledge base to support global change policy goals." *International Journal of Digital Earth* 13: 188-216. https://doi.org/10.1080/17538947.2018.1559367.

Ontobee. "Sustainable Development Goals Interface Ontology." Last accessed October 1, 2021. http://www.ontobee.org/ontology/SDGIO.

Petasis, Georgios, Vangelis Karkaletsis, Georgios Paliouras, Anastasia Krithara, and Elias Zavitsanos. 2011. "Ontology population and enrichment: State of the art." *Knowledge-driven multimedia information extraction and ontology evolution. Lecture Notes in Computer Science,* edited by Paliouras G., Spyropoulos C.D., Tsatsaronis G., 6050: 134-66. https://doi.org/10.1007/978-3-642-20795-2_6.

World Wide Web Consortium (W3C). "SKOS Simple Knowledge Organization System – Reference." Last accessed October 1, 2021. https://www.w3.org/TR/skos-ref erence/.

Zaino, Jennifer. "Ontology Plays a Part in United Nations Sustainable Development Goals Project." Last accessed October 1, 2021. https://www.dataversity.net/ontology-has-big-part-to-play-in-united-nations-sustainable-development-goals-project/#.

Zeng, Marcia Lei. 2019. "Interoperability." *Knowledge Organization*, 46, no. 2: 122-46.

Zeng, Marcia Lei and Lois Mai, Chan. 2015. "Semantic interoperability." *Encyclopedia of Library and Information Sciences 3rd edition*, edited by Marcia J. Bates and Mary Niles Maack. New York, NY: Dekker Encyclopedias, Taylor and Francis Group, 4645-62.

Zins, Chaims. 2007. "Conceptual approaches for defining data, information, and knowledge." *Journal of the American Society for Information Science and Technology* 58: 479-93.

Evaluation of LOINC semantic coverage for coding clinical observations related to environmental influenced diseases

Maria Teresa Chiaravalloti
Instituto di Informatica e Telematica,
Consiglio Nazionale delle Ricerche,
Cosenza, Italy

Abstract

Terminology standards ensure semantic interoperability by allowing efficient exchange and pool of data coming from many idiosyncratic sources. This is especially true when operating among multiple domains that are connected to each other through different kinds of semantic relationships. Environment and health are strictly related domains, as the first one produces a variety of effects on the second one. While terminologies exist for coding and classifying almost all the aspects of the healthcare process (actors, products, observations, etc.), a specific one focused on environmental related or caused health problems has not yet been developed. Nonetheless, concepts related to the link between the two domains can be found in different standards. Among them, Logical Observation Identifiers Names and Codes (LOINC®) is the most widely used for identifying parameters connected to clinical conditions. Considering this, the work presented in this paper aims to: 1) identify a subset of existing LOINC codes of clinical observations influenced or related to environmental factors; 2) search for environmental concepts not covered by LOINC in the UMLS Metathesaurus to see if they are covered and how they are expressed in other medical terminologies; and consequently 3) detect LOINC concept gaps related to the link between environment and health to be filled by proposing new LOINC terms for standardizing identified clinical parameters.

1.0 Introduction

Widespread adoption of terminology standards offers the promise of enabling efficient processing and storage of data that comes from many independent sources (McDonald 1997). Only if data carrying the same semantic meaning are identified by the same code, even if they are labeled with idiosyncratic names, can be pooled and merged. Existing standard terminologies cover multiple domains and they are often interrelated through meta-thesauri, mappings, transcodings, etc., so they can interoperate according to different types of semantic relationships. Multiple relations exist between *health* and *environment* domains, covering more than one aspect of each. However, the influence of environmental factors on human health is proved by clinical studies of different medical specialties (Mekonnen 2021, Nakayama 2021 and Stringer 2021, just to cite the most recent ones), and it is has been recently confirmed by the WHO Global Air Quality Guidelines (AQGs) (World Health Organization 2021). Compared to their last edition of 2005, AQGs reduce levels of key air pollutants that need to be controlled to keep air quality levels within limits which are not dangerous for human health. The guidelines, in fact, provide clear evidence of the damage inflicted by air pollution on the health of the population, at even lower concentrations than previously recommended.

Both *health* and *environment* domains are almost completely semantically covered by terminologies for coding and classifying different aspects of those fields. Nonetheless, to the best of our knowledge, a specific resource focused on environmental related or caused health problems has not yet been developed. However, concepts expressing the relation between the two domains can be found in different terminological standards. Among them, this work is focused on LOINC, which is the most widely used standard for identifying clinical observations connected to health conditions (not necessarily already resulted in problems or diseases). Investigating clinical parameters that can be influenced by environmental determinants is fundamental to monitor the trend and evolution of possible risk situations and consequently plan corrective actions. For this reason, this work aims to identify a subset of existing LOINC codes of clinical observations influenced or related to environmental factors, finding possible concept gaps and consequently proposing new LOINC terms to standardize them. The choice of LOINC is also due to the fact that it is an open and nimble standard, compared to many others of the clinical domain, and this makes it easy to quickly update.

2.0 Background

LOINC (McDonald 2003) is a universal code system for laboratory and clinical observations used in more than 164 countries to enable semantic interoperability. It was created in 1994 by the Regenstrief Institute, Inc, which continues as the standards development organization (SDO) responsible for its development. The current release, version 2.71 (August 2021), contains more than 96,000 concepts covering the full scope of laboratory testing (hematology, microbiology, etc.) and a broad range of clinical measurements (e.g. vital signs, EKG, patient reported outcomes, etc.). Based on formal naming conventions, LOINC also carries names for document titles (Frazier 2001) radiology reports (Vreeman 2005) and section headings (social history, objective, etc.).

At present, LOINC is used by many kinds of organizations, including large reference laboratories, healthcare organizations, insurance companies, regional health information networks and national standard bodies. Global LOINC adoption has been accelerated by an innovative approach to facilitating translation (Vreeman 2012). The official LOINC Italian version has been translated and it is currently mantained by the Institute of Informatics and Telematics of the National Council of Research. It has been adopted as national standard for identifying document type in the national Electronic Health Record, a plethora of clinical parameters in Health Level 7 (HL7) Implementation Guides and, specifically, for coding performed tests in laboratory reports.

LOINC codes are uniquely assigned numbers and LOINC names are defined 'fully specified' because they contain all the information needed to unambiguously identify a test, distinguishing it from others who might apparently seem identical. These names are given by the concatenation of six fundamental axes:

1. Component: the substance that is measured (e.g., sodium, glucose, etc.);
2. Property: the measurement type;
3. Timing: it distinguishes measurements made at a given time by those covering a time interval;
4. System: the type of sample on which the observation is performed;
5. Scale: the scale of measurement;
6. Method: the method used in test performing.

Studies about the application of specific subsets of LOINC codes have been carried out in multiple domains, such as workers' health in the workplace (Park 2021), social determinants of health (Watkins 2020) and, last but not least, the Covid-19 (Dong 2020), but there are not a lot of works about the use of LOINC for coding clinical observations in some way related to environmental factors, despite the fact that it is the most widely used standard for identifying measurements and parameters that allow the definition or monitoring of health problems and conditions. It is worth mentioning the Pan-Canadian REspiratory STandards INitiative for Electronic Health Records (PRESTINE), whose

> "goal is to recommend respiratory data elements and standards for use in electronic medical records across Canada that meet the needs of providers, administrators, researchers and policy makers to facilitate evidence-based clinical care, monitoring, surveillance, benchmarking and policy development" (Lougheed 2012, 117).

They identified clinical nomenclatures such as the Systematized Nomenclature of Medicine – Clinical Terms (SNOMED-CT®) and LOINC as standards currently available for clinical variables that are likely to be included in electronic medical records in primary care for diagnosis, management and patient education related to asthma and Chronic Obstructive Pulmonary Disease (COPD), and other respiratory conditions. Among the identified LOINC codes, there are many regarding clinical parameters influenced or determined by environmental aspects, thus making evident direct or indirect relationships between the two observed diseases and environmental determinants.

3.0 Materials and methods

The first step of this work was identifying existing LOINC terms that could be related, both broadly or narrowly, to environment conditioned factors, so as to evaluate the degree of semantic coverage of the standard in the target domain. They had been retrieved using entries of specific controlled terminologies as search keywords. Among the most used Knowledge Organization Systems (KOS) of the domain, we considered GEMET (General Multilingual Environmental Thesaurus),[1] AGROVOC (https://agrovoc.fao.org/ browse/agrovoc/en/), SWEET (Semantic Web for Earth and Environment Technology) Ontology,[2] Earth (which includes GEMET)[3] and INSPIRE Feature Concept Dictionary,[4] as references. Firstly, we used the search keyword *health* in all of them, as the aim was looking for how healthcare related concepts are represented in terminological resources primarily dedicated to the environmental domain. We retrieved 29 concepts from GEMET, 19 from AGROVOC, 4 from SWEET Ontology, 46 from Earth and 5 from INSPIRE. Then they were matched among each other so to create a list of unique concepts and avoid to double use them. This operation revealed that GEMET has the major part of its health-related concepts covered by the other considered terminological resources, while INSPIRE has no intersections with the other four vocabularies of the domain. The unique concepts of the list were 70 and they were used as searching keywords in LOINC to recall codes of those clinical observations or parameters that could be in some way related

1   European Environment Information and Observation Network (EIONET), "GEMET Thematic Listings," EIONET Portal, last accessed October 11, 2021, https://www.eio net.europa.eu/gemet/en/themes/.

2   National Center for Biomedical Ontology, "Semantic Web for Earth and Environment Technology Ontology," BioPortal, last accessed October 11, 2021, https://bio portal.bioontology.org/ontologies/SWEET/?p=classes&conceptid=root.

3   National Research Council – Institute of Polar Sciences (CNR-ISP), "EARTh – Environmental Applications Reference Thesaurus," last accessed October 11, 2021, https:// www.isp.cnr.it/index.php/en/earth.

4   European Commission, "INSPIRE feature concept dictionary," INSPIRE Knowledge Base, last accessed October 11, 2021, https://inspire.ec.europa.eu/featureconcept.

LOINC code	Component	Property	Scale	Time	System	Method	Class
89551-6	Consultation note	Find	Doc	Pt	[Setting]	Environmental health	DOC.ONTOLOGY
	A consultation note is generated by a provider upon request for an opinion or advice from another provider. Consultations may involve face-to-face time with the patient, telemedicine visits, or a second opinion on a diagnosis that does not involve interaction with a patient. A consultation note is typically sent to the referring provider when the consultation is completed.						
42561-1	Event description.environmental hazard &or safety	Find	Nom	Pt	^Patient	Observed.MERSTH	PATIENT SAFETY
	Normative Answer List LL415-1						
	AnswerCodeScoreAnswer ID						
	Body fluid exposureLA7393-7						
	Chemical exposureLA7394-5						
	Chemotherapy spillLA7395-2						
	Hazardous material spillLA7396-0						
	OtherLA46-8						
89552-4	Note	Find	Doc	Pt	[Setting]	Environmental health	DOC.ONTOLOGY
96336-3	Note	Find	Doc	Pt		Outpatient Environmental health	DOC.ONTOLOGY

Figure 1. Pertinent LOINC codes recalled by the searching keyword environmental health

to environmental aspects. Some concepts were excluded because too generic for a health-related terminology such as LOINC and they would have recalled a lot of noise (i.e *health, healthcare, health service, public health*). Obviously narrower keywords recalled the same LOINC codes of their correspondent broader terms, i.e. typing *environmental health hazard* we found one LOINC code, which is comprised in the ten codes recalled by *environmental health*. So narrower keywords were not considered, if they are covered by a related broader term. Then, we analyzed the LOINC codes recalled by each searching keywords in order to eliminate those not relevant for the purpose of this work. For example, *environmental health* recalled ten LOINC codes, but only four of them are pertinent for the considered domain (Figure 1).

As 25 out of 70 searching keywords of our unique concepts list did not recall any LOINC code and also many of the retrieved ones are not relevant for the aims of the study, we decided to search in LOINC using the upper categories of the considered KOSs as keywords. Also in this case we excluded extremely generic words such as *administration* in GEMET. This choice created more noise, but it also gave the chance to retrieve more LOINC codes, which are logically related to the "health and environment" domain even if they do not contain those two words in the code description string. Consequently, more analysis work has been required on the recalled codes to fully understand the context of use. At the end of this process, we have identified a subset of 476 LOINC codes of clinical observations in some way related to environmental influence.

As second step, in order to verify if there could be possible LOINC concept gaps to fill by submitting requests for the creation of new codes to the LOINC Committee, we searched the UMLS (Unified Medical Language System) Metathesaurus[5] for those concepts not semantically covered by LOINC,

---

5 «The Metathesaurus is the biggest component of the UMLS. It is a large biomedical thesaurus that is organized by concept, or meaning, and it links similar names for the

but that are representative of the reference domain (i.e *soil*). This activity will allow a cross-checking of the reference concepts in multiple KOSs, especially those coding problems and diseases such as the International Classification of Diseases (ICD). Obtained results will support the need to create new LOINC codes of clinical observations immediately referable to diseases determined by environmental factors encoded in other medical terminologies.

4.0 Results

The LOINC codes retrieved according to the methodology described in the previous paragraph are quite representative enough of the target domain of this work. They are not systematically organized under a common and dedicated Class tag within the standard, but if considered together they actually well represent those clinical parameters in some way related or influenced by environmental factors.

The keywords that matched more pertinent LOINC codes are *air* and *water*, with 195 and 156 retrieved concepts respectively. That is because they are values of the LOINC System axis, expressing the means in which polluting substances can be measured. The majority of these codes belong to the Toxicology LOINC Class and report observations about substances in water or air that could in some way affect human health. Figure 2 and Figure 3 show the descriptions associated to the LOINC codes retrieved, where we underlined in red the parts containing the effects on health of the measured substances, in *air* and in *water* respectively.

Also the LOINC codes retrieved using the keyword *environmental health* belong to the Toxicology Class. They mainly search for the presence of toxic or harmful substances for the human body or metals or volatile organic compounds, both present in nature and by chemical composition. Most of them are panels required by the Center for Environmental Health (CEH),[6] which is an American organization aiming at protecting people from harmful chemicals in air, food, water and in everyday products. A panel in LOINC is a list of clinical observations usually made together. For example, in Figure 4 there is a panel about air contaminants that could be detected in a house.

---

same concept from nearly 200 different vocabularies. The Metathesaurus also identifies useful relationships between concepts and preserves the meanings, concept names, and relationships from each vocabulary». Cfr. National Institute of Health (NIH), "UMLS Metathesaurus," National Library of Medicine (NLM), last reviewed April 12, 2016, https://www.nlm.nih.gov/research/umls/knowledge_sources/metathesaurus/index.html.

6   "Center for Environmental Health," Wikipedia, last accessed October 11, 2021, https://en.wikipedia.org/wiki/Center_for_Environmental_Health.

38624-3	1,1,2-Trichloroethane [Mass/volume] in Air
	1,1,2-Trichloroethane, or 1,1,2-TCA, is an organochloride solvent with the molecular formula C2H3Cl3. It is a colorless, sweet-smelling liquid that does not dissolve in water, but is soluble in most organic solvents. It is an isomer of 1,1,1-trichloroethane. It is used as a solvent and as an intermediate in the synthesis of 1,1-dichloroethane. 1,1,2-TCA is a central nervous system depressant and inhalation of vapors may cause dizziness, drowsiness, headache, nausea, shortness of breath, unconsciousness, or cancer. Trichloroethane may be harmful by inhalation, ingestion and skin contact. It is a respiratory and eye irritant. Although no definitive studies currently exist, trichlorethane should be treated as a potential carcinogen since laboratory evidence suggests that low molecular weight chlorinated hydrocarbons may be carcinogenic
38625-0	Trichloroethylene [Mass/volume] in Air
	A highly volatile inhalation anesthetic used mainly in short surgical procedures where light anesthesia with good analgesia is required. It is also used as an industrial solvent. Prolonged exposure to high concentrations of the vapor can lead to cardiotoxicity and neurological impairment.
38594-8	Epichlorohydrin [Mass/volume] in Air
	A chlorinated epoxy compound used as an industrial solvent. It is a strong skin irritant and carcinogen
38693-8	Ethyl acrylate [Mass/volume] in Air
	Ethyl acrylate is an organic compound primarily used in the preparation of various polymers. It is a clear liquid with an acrid penetrating odor. The human nose is capable of detecting this odor at a thousand times lower concentration than is considered harmful if continuously exposed for some period of time.

Figure 2. LOINC codes and descriptions of effects on human health of some substances measured in the air

62534-3	PhenX - environmental exposures - air contaminants in the home environment protocol 061101
Indent63790-0	In the past year has there been a major renovation to this house or apartment, such as adding a room, putting up or taking down a wall, replacing windows, or refinishing floors?
Indent63791-8	What type of renovation?
Indent63792-6	When was the last renovation?
Indent63793-4	Within the last six months were rugs, drapes or furniture professionally cleaned?
Indent63794-2	Were they cleaned inside the house?
Indent63795-9	When were they cleaned?
Indent63796-7	What items were cleaned?
Indent63797-5	In the past year, was the inside of this house or apartment painted?
Indent63798-3	When was the last time?
Indent63799-1	On how many rooms?
Indent63845-2	In the past year were new carpets or rugs installed?
Indent63846-0	When was the installation?
Indent63847-8	On how many rooms?
Indent63800-7	Did you go to the dry cleaners during the past week?
Indent63801-5	Did you bring home any items from the cleaners that were dry-cleaned during the past week?
Indent63802-3	Specify air contaminant in the home [RIOPA]
Indent63808-0	Which cleaning solutions [RIOPA]
Indent63803-1	Have you used Paints or solvents (paint thinners and removers, typewriter corrective fluids) or has someone used near you in the last 48 hours?
Indent63804-9	Did you handle Paints or solvents (paint thinners and removers, typewriter corrective fluids) yourself? If Yes, for how long?
Indent63805-6	How long did you handle Paints or solvents (paint thinners and removers, typewriter corrective fluids) yourself?

Figure 3. LOINC codes and descriptions of effects on human health of some substances measured in the water

LOINC was born to provide unique identifiers for 'what' is measured, so if an observation is a question and its result is an answer, LOINC codes are created for the question. The answer can be coded using other standardized terminologies, such as SNOMED CT (Systematized Nomenclature of Medicine – Clinical Terms) (https://www.snomed.org/). Nonetheless, as

"Across many domains, the meaning of a particular observation can be best understood in the context of the set of possible answers (result values)" (Regenstrief Institute 2021, 114).

38334-9	**1,1,2-Trichloroethane [Mass/volume] in Water** 1,1,2-Trichloroethane, or 1,1,2-TCA, is an organochloride solvent with the molecular formula C2H3Cl3. It is a colorless, sweet-smelling liquid that does not dissolve in water, but is soluble in most organic solvents. It is an isomer of 1,1,1-trichloroethane. It is used as a solvent and as an intermediate in the synthesis of 1,1-dichloroethane. 1,1,2-TCA is a central nervous system depressant and inhalation of vapors may cause dizziness, drowsiness, headache, nausea, shortness of breath, unconsciousness, or cancer. Trichloroethane may be harmful by inhalation, ingestion and skin contact. It is a respiratory and eye irritant. Although no definitive studies currently exist, trichloroethane should be treated as a potential carcinogen since laboratory evidence suggests that low molecular weight chlorinated hydrocarbons may be carcinogenic
38292-9	**1,1-Dichloroethylene [Mass/volume] in Water** 1,1-Dichloroethene, commonly called 1,1-dichloroethylene or 1,1-DCE, is an organochloride with the molecular formula C2H2Cl2. It is a colorless liquid with a sharp odor. Like most chlorocarbons, it is poorly soluble in water, but soluble in organic solvents. 1,1-DCE was the precursor to the original cling-wrap for food, but this application has been phased out. The health effects from exposure to 1,1-DCE are primarily on the central nervous system, including symptoms of sedation, inebriation, convulsions, spasms, and unconsciousness at high concentrations
38332-3	**1,2,4-Trichlorobenzene [Mass/volume] in Water** 1,2,4-Trichlorobenzene is an organic compound used as a solvent, and is one of the best known solvents used to dissolve fullerenes and pentacene. It is a benzene derivative with three chlorine atoms substitutents, in the 1, 2 and 4 positions of the benzene ring. Trichlorobenzene (TCB) may refer to any of three isomeric chlorinated derivatives of benzene with the molecular formula C6H3Cl3: - 1,2,3-Trichlorobenzene - 1,2,4-Trichlorobenzene - 1,3,5-Trichlorobenzene
38288-7	**1,2-Dibromo-3-Chloropropane [Mass/volume] in Water** 1,2-Dibromo-3-chloropropane, (dibromochloropropane) better known as DBCP, is the active ingredient in the nematicide Nemagon, also known as Fumazone. It is a soil fumigant formerly used in American agriculture. In mammals it causes male sterility at high levels of exposure.

Figure 4. LOINC panel 62534-3 about air contaminants in the home environment

63802-3	Specify air contaminant in the home [RIOPA]
	Paints or solvents, paint thinners and removers, typewriter corrective fluids
	Glues and adhesives
	Gasoline lawn mower
	Chain saw or other gasoline equipment
	Sander and/or saw
	Pesticides sprayed
	Vacuuming
	Sweeping indoors
	Dusting
	Cleaning solutions, including household cleaners and chemicals
	Gardening
	Woodworking
	Metal working or welding
	Broiling, smoking, grilling or frying inside the house
	Broiling, smoking, grilling or frying outside the house

Figure 5. LOINC Answers List of the code 63802-3 specifying home air contaminants

the LOINC Committee decided to create LOINC answer lists for certain types of observations. They can be *example*, *preferred* or *normative*, which bind the choice of that LOINC code to the use of its values. For example, LOINC Answers (LA) Lists in Figure 5 show which air contaminants of the home can be specified as results of the LOINC observation 63802-3.

Exploring other results, the concept *disaster* retrieved only two pertinent LOINC codes (Figure 6). One (LOINC 66886-3) investigates if the patient has ever been involved in a natural disaster and it is part of Lifetime Trauma and Victimization History (LTVH) survey, which aims at inquiring about

66886-3	Have you ever been involved in a natural disaster, such as a tornado, hurricane, flood, or earthquake [LTVH]	
69463-8	Suspected intentional or unintentional disaster NEMSIS	
	Biologic agent	4507001
	Building failure	4507003
	Chemical agent	4507005
	Explosive device	4507007
	Fire	4507009
	Hostage event	4507011
	Mass gathering	4507013
	Mass illness	4507015
	Nuclear agent	4507017
	Radioactive device	4507019
	Secondary destructive device	4507021
	Shooting/sniper	4507023
	Vehicular	4507025
	Weather	4507027

Figure 6. LOINC codes retrieved by the keyword disaster *and LA list of the second one*

	Action taken based on risk assessment meeting	
92036-3		PUBLICHEALTH
91022-4	Date and time of risk assessment	PUBLICHEALTH
91880-5	Date and time of risk assessment meeting	PUBLICHEALTH
28191-5	**Poisoning risk [CCC]**	
	Exposure to or ingestion of dangerous products	
91578-5	**Probability of an event occurring**	
	The level of probability (e.g. very unlikely, unlikely, likely, highly likely, almost certain) that the event will occur or re-occur. An event that is very likely (almost certain) is one that will occur (or re-occur) in most circumstances when the activity is undertaken. An event that is very unlikely (rare) may happen only in exceptional circumstances when the activity is undertaken.	
74019-1	**Probability of developing disease assessed [Likelihood]**	
	The estimated likelihood of a patient developing a disease. Probability may be based a number of factors, including family history (pedigree) analysis, genetic algorithm results, pre- or post-test results, and general population data for disease prevalence.	
91023-2	**Reason for risk assessment**	
	The reason the risk assessment is conducted. The reason may simply be the incident or event name.	PUBLICHEALTH
89933-6	**Relative risk of developing disease assessed**	
	Relative risk or risk ratio (RR) of developing disease is the ratio of the probability of developing a disease in an exposed group to the probability of the event occurring in a non-exposed group.	
71482-4	risk assessment document	

Figure 7. LOINC codes retrieved by the keyword risk *and their descriptions*

30 traumas and victimization experiences people can undergo during their lifetime (Widom 2005). The other aims to code the type of disaster the patient experienced, giving the chance to choose among some in a normative LA list, and it was created upon request of the National Emergency Medical Services Information System (NEMSIS) (https://nemsis.org/).

The keyword *risk* gave back 12 pertinent LOINC codes out of 492 retrieved records (Figure 7). Among them, there are some to be noticed: *Poisoning risk* (LOINC 28191-5), which registers the patient exposure to or ingestion of dangerous products; *Probability of developing disease assessed* (LOINC 74019-1), which is estimated on a number of factors, including family history (pedigree) analysis, genetic algorithm results, pre- or post-test results, and general population data for disease prevalence, but not on risks related to environmental factors, both natural or determined by the place where the patient lives or works. So, in this case, the code could work for the purpose

75274-1	**Characteristics of residence**
	The patient's home-setting circumstances reflecting the status of the accommodations, living situation, and environment
63839-5	**Do you have wall-to-wall carpeting in your home [FEAS]**
	excluding the bathroom
63833-8	**Floor material in your workspace if other [FEAS]**
63837-9	Has there been renovation in your home during the past 12 months [FEAS]
63836-1	Has there been renovation or repairs in your home due to moisture damage [FEAS]
63831-2	Has there been renovation or repairs in your workspace because of moisture damage [FEAS]
71802-3	**Housing status**
	Describes patients living arrangement
63840-3	**Is the floor material in your home plastic or vinyl [FEAS]**
63842-9	**Is the wall material in your home plastic [FEAS]**
63841-1	**Is the wall material in your home textile, cloth, jute, etc [FEAS]**
63835-3	**Is the wall material of your work space plastic [FEAS]**
63834-6	**Is the wall material of your work space textile, cloth, jute, etc [FEAS]**
96778-6	**Problems with place where you live**
	Pests such as bugs, ants, or mice
	Mold
	Lead paint or pipes
	Lack of heat
	Oven or stove not working

Figure 8. Some of the LOINC codes retrieved by the keyword pollution

of registering a clinical observation related to the influence of environment on health, but its description needs to be updated. Similarly, for the LOINC 71482-4 *Risk assessment document*, though it was originally created for use within the Cardiac Imaging Report CDA Implementation Guide developed by IHE (Integrating the Healthcare Enterprise), the description says it is not limited in use to it, but the chance to refer it to environmental related risks is not made explicit. They are all codes belonging to the LOINC Class *Public health* and that is indicative of the scope and purpose for which they were created.

Last but not least, it is worth mentioning the concept *pollution*. Looking at the retrieved LOINC codes (Figure 8), they are mostly related to the house environment or workspace, as they investigate aspects such as the floor and wall material, the presence of wall-to-wall carpeting and other general problems (i.e. pests, mold, lead paint or pipes, lack of heat, oven or stove not working). There are no codes related to the pollution that can be caused by natural factors or human activities (such as industries) and could cause effects on human health.

Despite the numerous and pertinent LOINC codes retrieved through the described keywords, many concepts of the upper categories of the considered KOSs, which are representative of the reference domain, did not find any correspondence in the standard. Among them, those for which it would be appropriate to have a representation in LOINC, because in some way relatable to environmental factors influencing human health, are: *agriculture, biology, climate, energy, forestry, natural areas, ecosystems, natural dynamics, vibrations, soil, transport, urban environment, urban stress*. The second step of this work, as described in the previous paragraph, was searching for these high-level concepts in the UMLS Metathesaurus, in order to see if and how they are represented in other medical terminologies and so determining clinical semantic areas that would need to be covered by LOINC. In particular, searching results belonging to KOSs that encode diseases and problems were considered, as it is consequential that clinical observations or parameters are mostly defined in relation to them.

Using the concept *agriculture* as search keyword in the UMLS Metathesaurus Browser[7] and looking for vocabulary codes,[8] we obtained 49 results. Among them, one is relevant for the purpose of this study: *occupational exposure to toxic agents in agriculture*, which is encoded in ICD10[9] (Z57.4), ICD10CM[10] (Z57.4), ICD10AM[11] (Z57.4) and MEDCIN[12] (129080). This concept could be formalized in LOINC similarly to codes that evaluate occupational risks due to the presence harmful substances, in this case specifically related to the agricultural work.

Searching for *biology* gives back 155 results, but none of them is useful in identifying clinical conditions determined by environmental factors, because retrieved codes are mainly of biology subcategories so the term in this form is surely inclusive of concepts of interest but too high level for our aims.

Typing the word *climate*, 56 resulting codes are recalled. Most of them are for climate types, but some can be considered as in some way expressing the cause effect relationship between climate changes (intended both as a

---

7   National Institute of Health (NIH), "UMLS Metathesaurus Browser," National Library of Medicine, last accessed October 11, 2021, https://www.nlm.nih.gov/research/umls/knowledge_sources/metathesaurus/index.html.

8   It is also possible to search for UMLS concepts. In this case, codes belonging to different vocabularies but carrying the same semantic meaning are grouped together under the same UMLS CUI (Concept Unique Identifier). For example, searching for agriculture will retrieve 19 different CUIs against 49 vocabulary codes, meaning that those 49 carry 19 different concepts, some of which are repeated in more than one code system so they can be conceptually gathered together.

9   World Health Organization (WHO), "International Classification of Diseases 10th revision," last accessed October 11, 2021, https://www.who.int/standards/classifications/classification-of-diseases.

10   The Centers for Disease Control (CDC), "International Classification of Diseases, Tenth Revision, Clinical Modification (ICD-10-CM)," Centers for Diseases Control and Prevention, last accessed October 11, 2021, https://www.cdc.gov/nchs/icd/icd10cm.htm.

11   Independent Hospital Pricing Authority (IHPA), "International Classification of Diseases 10th revision Australian Modification," last accessed October 11, 2021, https://www.ihpa.gov.au/what-we-do/icd-10-am-achi-acs-current-edition.

12   Medicomp Systems, "The Medicomp MEDCIN Engine," last accessed October 11, 2021, https://medicomp.com/medcin/.

change in climate due to a change in geographical area and as a major climate change in general) and health status. They are:

- *recently traveled to or resided in tropical climate* (MEDCIN 3152);
- *recently traveled to or resided in desert climate* (MEDCIN 3153);
- *recently traveled to or resided in temperate climate* (MEDCIN 3150);
- *recently traveled to or resided in subtropical climate* (MEDCIN 3151);
- *traveled to or resided in climate which is:* (MEDCIN 3149);
- *climate change* (MEDLINEPLUS 5144) (https://medlineplus.gov/), which is a subcategory of *Poisoning, Toxicology, Environmental health;*
- *climatotherapy* (MESH[13] D013790), which consists into the temporary or permanent relocation of a patient in a place with a climate that can help the recovery or improvement of a pathological condition.

The 1,628 codes recalled by the searching keyword *energy* are mostly referred to therapies and treatments that involve the use of some form of electric energy, to energy intended as physical vigor or to the thermal energy potential enclosed in a food. While only the code 242720004 *Accident caused by heat energy release from machine* from SNOMED CT refers to accidental harms caused by heat energy, therefore considering the concept from a negative point of view respect to the other uses.

The concept *forestry* does not have any code from the medical terminologies in the UMLS that could be related to some influence on human health, but it just recalls 24 codes mostly about forestry as a place or as a job sector. Similarly, the queries using the concepts *natural areas* and *natural dynamics* yielded no results, as probably they are too high level for our purposes. While using *ecosystems* as search keyword gives only 2 results, meaning that it has not been semantically explored in any medical terminology of the UMLS yet.

Useful to be encoded in LOINC would be the concept conveyed by the MedDRA French Editon[14] code 10030023 *Exposition professionnelle aux vibrations*, which is among the 6 codes recalled by the search term *vibrations*.

The search keyword *soil* retrieved the highest number of meaningful resulting codes. Many of them refer to different types of *soil bacteria* and *soil fungi* and come from the NCBI (National Center for Biotechnology) Taxonomy.[15] There are two codes for *soil pullutants*, one of which specifying if they

---

13 National Center for Biotechnology Information (NCBI), "Medical Subject Headings," last accessed October 11, 2021, https://www.ncbi.nlm.nih.gov/mesh/.

14 International Council for Harmonisation of Technical Requirements for Pharmaceuticals for Human Use (ICH), "MedDRA Support Documentation, Medical Dictionary for Regulatory Activities," last accessed October 11, 2021, https://www.meddra. org/how-to-use/support-documentation/french.

15 National Center for Biotechnology Information (NCBI), "Taxonomy," last accessed October 11, 2021, https://www.ncbi.nlm.nih.gov/guide/taxonomy/.

are *radioactive*, from MESH (D012989 and D012990, respectively). Similarly, a number of codes from different medical terminologies encodes the concept *soil pollution*:

- A-A3040 of SNMI;[16]
- 2733-6435 of CSP;[17]
- 102417007 of SNOMEDCT_US;[18]
- 0000038184 of CHV (https://consumerhealthvocab.org/).

Other than these codes, which are essentially used to identify organisms, substances or a certain condition of the soil, there are those specifically expressing a cause-effect relation between a polluted or contaminated soil and the health status. Among them, it is also possible to make a distinction between codes expressing an exposure or contact because of occupational or non-occupation reasons:

- *Exposure to polluted soil* (SNOMEDCT_US 102432000);
- *Exposure to polluted soil, occupational* (SNOMEDCT_US 102433005);
- *Exposure to polluted soil, non-occupational* (SNOMEDCT_US 102434004);
- *Contact with and (suspected) exposure to soil pollution* (ICD10CM Z77.112);
- *Exposure to polluted soil, NOS* (SNMI A-A3240);
- *Exposure to polluted soil, occupational* (SNMI A-A3241);
- *Exposure to polluted soil, non-occupational* (SNMI A-A3242);
- *Exposure to polluted soil* (MedDRA 10063603) (https://www.meddra.org/);
- *Non-occupational exposure to polluted soil* (MedDRA 10056402);
- *Exposure to soil pollution* (ICD10AM Z58.3);
- *Exposure to soil pollution* (MEDCIN 129110);
- *Exposure to possibly contaminated soil* (MEDCIN 3293).

The concepts vehicled by these codes are not covered by LOINC at all, so they are meaningful candidates for formulating a new term request submission to the LOINC Committee.

---

16 SNOMED International 1998, which «has been superseded by the Systematized Nomenclature of Medicine – Reference Terminology (SNOMED RT®). SNOMED RT has been merged with Clinical Terms Version 3 (CTV3), also known as Read Codes, resulting in the creation of SNOMED Clinical Terms (SNOMED CT®)». National Library of Medicine (NLM), "SNMI (SNOMED Intl 1998) – Synopsis," last accessed October 11, 2021, https://www.nlm.nih.gov/research/umls/sourcereleasedocs/current/SNMI/index.html.

17 National Institute of Health (NIH), "CSP (CRISP Thesaurus) – Synopsis," National Library of Medicine, last accessed October 11, 2021, https://www.nlm.nih.gov/research/umls/sourcereleasedocs/current/CSP/index.html.

18 National Institute of Health (NIH), "SNOMED CT United States Edition," National Library of Medicine, last accessed October 11, 2021, https://www.nlm.nih.gov/health it/snomedct/us_edition.html.

5.0 Discussion and conclusions

Standardized KOSs need continuous updates if they want to keep up with the ongoing discoveries and evolution of the scientific world and, moreover, to maintain their usefulness as representation and access tools to the knowledge of a specific domain. Cross references among them are vital as they avoid leaving them as single islands and create a complete representation of multiple and interacting aspects of different domains. The links between environment and health are increasing hand in hand with the influence that environmental factors have on the development and progression of certain diseases. Nonetheless, this semantic area is not yet fully represented in standardized medical terminologies especially for what pertains to observable clinical parameters (as trigger factors, consequences, indicators, etc.) related to a disease. This paper aimed to investigate existing LOINC codes for this semantic area, then searching concepts not covered in the UMLS Metathesaurus to see if they are covered and how they are expressed in other medical terminologies and consequently detecting LOINC concept gaps related to the link between environment and health to be filled by proposing new LOINC terms for standardizing identified clinical parameters.

The study gave good results in terms of LOINC evaluation as both gold standard for identifying clinical observations also in the environmental domain and for detecting semantic area not covered by the standard and therefore susceptible to being modeled by requesting the creation of new *ad hoc* LOINC codes. Despite the fact that it is the most widely used standard for uniquely identifying laboratory and clinical observations, it has not fully developed sections or health related domains that need to be addressed more completely. On the other hand, compared to other medical terminologies, LOINC has on its side that it has an open, rapid and pragmatic updating process as it is based on user requests and therefore on the real coding needs connected to new scientific discoveries or evolutions. So, the results of this work could concretely contribute to the development of the standard and they will therefore be formalized according to the LOINC naming conventions to be submitted to the LOINC Committee, that will evaluate the creation of new LOINC codes for the identified clinical observations. Finally, based on the results of this study, the LOINC Committee could evaluate the opportunity to create a LOINC Class entirely dedicated to clinical observations related to environmental factors.

Although this work could have been difficult to approach due to the many facets that the concept of *environment* applied to the world of *health* can assume, the defined working methodology has brought significant results both in terms of numbers and quality. However, it is possible to trace some limits of the study, which can actually be considered as starting points for future refinements or in-depth studies. Among them, the chance to consider, in order to create the subset of searching keywords: I) more environmental vocabularies; II) not only high-level concepts, but also their subordinate, if significative; III) clinical concepts not only traceable through the term *health*. Moreover, the UMLS Metathesaurus is surely an important source of information as it allows to access multiple terminologies speedly and simultaneously, but it would be further useful to use it just as a starting point and then investigate each concept in its own vocabulary in order to consider its position in a hierarchical tree, possible meaningful branches, etc.

In the future, this work could be expanded through the collaboration with a medical expert for double-checking the identified concept gaps and evaluating if there are clinical conditions or problems that are in some way related to the environment influence but that are not yet encoded in any medical terminology.

In conclusion, the relationships between the environmental and medical domains are proven by various studies and it is of fundamental importance to make them explicit through measurable parameters as they allow them to be monitored and corrective actions taken in both sectors, for the benefit of both the environment and human health. To pursue this aim, standard terminologies ensure semantic interoperability by uniquely identifying concepts and creating semantic relationships among them using tools like the UMLS Metathesaurus. Open, nimble and pragmatic vocabularies like LOINC offer the chance to model the topic of this paper, which crosses multiple domains, according to the needs of the final users and without getting stuck in the rigidity of the branches of a classification system. Representing the knowledge of transversal domains is a non-trivial task, which has to be approached from multiple points of view, but which is necessary to have a shared identification of new arising concepts.

References

Centers for Disease Control (CDC). "International Classification of Diseases, Tenth Revision, Clinical Modification (ICD-10-CM)." Centers for Diseases Control and Prevention. Last accessed October 11, 2021. https://www.cdc.gov/nchs/icd/icd10cm.htm.

"Center for Environmental Health." Wikipedia. Last accessed October 11, 2021. https://en.wikipedia.org/wiki/Center_for_Environmental_Health.

Dong, Xiao, Jianfu Li, Ekin Soysal, Jiang Bian, Scott L. DuVall, Elizabeth Hanchrow, Hongfang Liu, Kristine E. Lynch, Michael Matheny, Karthik Natarajan, et al. 2020. "COVID-19 TestNorm: A tool to normalize COVID-19 testing names to LOINC codes." *Journal of American Medical Informatics Association* 27, no. 9: 1437-42.

European Commission. "INSPIRE feature concept dictionary." INSPIRE Knowledge Base. Last accessed October 11, 2021. https://inspire.ec.europa.eu/featureconcept.

European Environment Information and Observation Network (EIONET). "GE-MET Thematic Listings." EIONET Portal. Last accessed October 11, 2021. https://www.eionet.europa.eu/gemet/en/themes/.

Frazier, Pavla, Angelo Rossi-Mori, Robert H. Dolin, Liora Alschuler, and Stanley M. Huff. 2001. "The creation of an ontology of clinical document names." *Studies in Health Technologies and Informatics* 84: 94-8.

Independent Hospital Pricing Authority (IHPA). "International Classification of Diseases 10th revision Australian Modification." Last accessed October 11, 2021. https://www.ihpa.gov.au/what-we-do/icd-10-am-achi-acs-current-edition.

International Council for Harmonisation of Technical Requirements for Pharmaceuticals for Human Use (ICH). "MedDRA Support Documentation. Medical Dictionary for Regulatory Activities." Last accessed October 11, 2021. https://www.meddra.org/how-to-use/support-documentation/french.

Lougheed, M. Diane, Janice Minard, Shari Dworkin, Mary-Ann Juurlink, Walley J. Temple, Teresa To, Marc Kohen, Anne Van Dam, and Louis-Philippe Boulet. 2012. "Pan-Canadian REspiratory STandards INitiative for Electronic Health Records (PRESTINE): 2011 National Forum Proceedings." *Canadian Respiratory Journal* 19: 117-26. https://doi.org/10.1155/2012/870357.

McDonald, Clement J. 1997. "The barriers to electronic medical record systems and how to overcome them." *Journal of the American Medical Informatics Association* 4: 213-21. https://doi.org/10.1136/jamia.1997.0040213.

McDonald, Clement J., Stanley M. Huff, Jeffrey G. Suico, Gilbert Hill, Dennis Leavelle, Raymond Aller, Arden Forrey, Kathy Mercer, Georges DeMoor, et al. 2003. "LOINC, a universal standard for identifying laboratory observations: a 5-year update." *Clinical Chemistry* 49: 624-33. https://doi.org/10.1373/49.4.624.

Medicomp Systems. "The Medicomp MEDCIN Engine." Last accessed October 11, 2021. https://medicomp.com/medcin/.

Mekonnen, Tesfaye H., Awrajaw Dessie, and Amensisa H. Tesfaye. 2021. "Respiratory symptoms related to flour dust exposure are significantly high among small and medium scale flour mill workers in Ethiopia: a comparative cross-sectional survey." *Environmental health and preventive medicine* 26, 1: 96. https://doi.org/10.1186/s12199-021-01019-y.

Nakayama, Kazumi, Yoshihide Sugita, Hideaki Tezuka, Koichi Takizawa, and Shoichi Muto. 2021. "Evaluation of the effect of radioactivity to the environment in wooden houses within the evacuation area at Fukushima." *Journal of radiological protection: official journal of the Society for Radiological Protection* 41, no. 4. https://doi.org/10.1088/1361-6498/ac2b8d.

National Center for Biotechnology Information (NCBI). "Medical Subject Headings." Last accessed October 11, 2021. https://www.ncbi.nlm.nih.gov/mesh/.

National Center for Biotechnology Information (NCBI). "Taxonomy." Last accessed October 11, 2021. https://www.ncbi.nlm.nih.gov/guide/taxonomy/.

National Institute of Health (NIH). "CSP (CRISP Thesaurus) – Synopsis." National Library of Medicine. Last accessed October 11, 2021. https://www.nlm.nih.gov/research/umls/sourcereleasedocs/current/CSP/index.html.

National Institute of Health (NIH). "UMLS Metathesaurus." National Library of Medicine. Last reviewed April 12, 2016. https://www.nlm.nih.gov/research/umls/knowledge_sources/metathesaurus/index.html.

National Institute of Health (NIH). "UMLS Metathesaurus Browser." National Library of Medicine. Last accessed October 11, 2021. https://www.nlm.nih.gov/research/umls/knowledge_sources/metathesaurus/index.html.

National Library of Medicine (NLM). "SNMI (SNOMED Intl 1998) – Synopsis." Last accessed October 11, 2021. https://www.nlm.nih.gov/research/umls/sourcereleasedocs/current/SNMI/index.html.

National Research Council – Institute of Polar Sciences (CNR-ISP). "EARTh – Environmental Applications Reference Thesaurus." Last accessed October 11, 2021. https://www.isp.cnr.it/index.php/en/earth.

Park, Hyun S., Kwang I. Kim, Ho-Young Chung, Sungmoon Jeong, Jae Y. Soh, Young H. Hyun, and Hwa S. Kim. 2021. "A Worker-Centered Personal Health Record App for Workplace Health Promotion Using National Health Care Data Sets: Design and Development Study." *JMIR medical informatics* 9, 8:e29184. https://doi.org/10.2196/29184.

Regenstrief Institute (RI). 2021. "LOINC Users' guide." Last accessed October 11, 2021. https://loinc.org/downloads/loinc-table/.

Stringer, Leandra, Tina L. Ly, Nicolas V. Moreno, Christopher Hewitt, Michael Haan, and Nicholas Power. 2021. "Assessing geographic and industry-related trends in bladder cancer in Ontario: A population-based study." *Canadian Urological Association journal*. https://doi.org/10.5489/cuaj.7263.

Vreeman, Daniel J., and Clement J.McDonald. 2005. "Automated mapping of local radiology terms to LOINC." In *AMIA Annual Symposium Proceedings, 22-26 October 2005 Washington, DC,* 769-73.

Vreeman, Daniel J., Maria T. Chiaravalloti, John Hook, and Clement J. McDonald. 2012. "Enabling international adoption of LOINC through translation." *Journal of Biomedical Informatics* 45, no. 4: 667-73. https://doi.org/10.1016/j.jbi.2012.01.005.

Watkins, Michael, Benjamin Viernes, Viet Nguyen, Leonardo Rojas Mezarina, Javier Silva Valencia, and Damian Borbolla. 2021. "Translating Social Determinants of Health into Standardized Clinical Entities." *Studies in health technology and informatics* 270: 474-78. https://doi.org/10.3233/SHTI200205.

Widom, Cathy S., Mary A. Dutton, Sally J. Czaja, and Kimberly A. DuMont. 2005. "Development and validation of a new instrument to assess lifetime trauma and victimization history." *Journal of Traumatic Stress* 18, no. 5: 519-31. https://doi.org/10.1002/jts.20060.

World Health Organization (WHO). "International Classification of Diseases 10th revision." Last accessed October 11, 2021. https://www.who.int/standards/classifications/classification-of-diseases.

World Health Organization (WHO). "WHO global air quality guidelines: particulate matter (PM2.5 and PM10), ozone, nitrogen dioxide, sulfur dioxide and carbon monoxide." Last accessed October 11, 2021. https://apps.who.int/iris/handle/10665/345329. License: CC BY-NC-SA 3.0 IGO.

Exploring semantic connections through a thesaurus in the earth observation domain[1]

Erika Pasceri
University of Calabria, Italy

Claudia Lanza
University of Calabria, Italy

Anna Perri
University of Calabria, Italy

Abstract

This paper presents the creation of a thesaurus for the Earth Observation (EO) systems domain which also includes terms representing the potential clinical effects caused by the pollutants. The goal is to provide a terminological means of support for professionals and public audiences in understanding specialized terms and their semantic connections with other representatives of this area of study (Bernier-Colborne 2012). The paper presents, in the first instance, the steps carried out in compiling the source corpus from which the specialized information about the contaminant substances has been retrieved; subsequently, the work will address the extraction of the specific terminology with the aid of pre-trained semantic software as well as the identification of a sub-methodological procedure to improve term configuration, i.e., semantic relation configuration by isolating the domain-specific connective verbal constructions (Auger 2008) and exploiting the textual contextualization to broaden the conceptual system and enhance the semantic accuracy.

1.0 Introduction

To achieve interoperability and avoid semantic ambiguity, the healthcare community has largely identified the need to develop standard codes for commonly used concepts and terms in healthcare delivery. Code-based vocabularies, terminologies, and classifications have been thoroughly developed: the standardization task in this area involves the development (or the constant update) of standard code-sets for generally used concepts, terms,

---

[1]   Authors have equally contributed to this work, however Erika Pasceri particularly focused on "Introduction", "The Method", "Corpus compilation" and "Text Processing"; Claudia Lanza focused on "Thesaurus construction", "Semantic interoperability" and "Results"; Anna Perri focused on "Discussion" and "Conclusion".

disease names, procedures, entity names, laboratory tests, observations, devices, clinical findings, pharmaceutical products, organisms, etc. The most relevant application of this kind of information is to assist in the meaningful interpretation of healthcare information exchanged across healthcare systems (Sinha et al. 2012). The more information we share, the more knowledge we achieve and that is particularly true in using knowledge organization systems that embed cross information useful for those systems created to support decision making processes.

The thesaurus is one of the most common Knowledge Organization Systems (KOSs) and it has been designed to improve information retrieval and to reduce the ambiguity of natural language in describing items for searching goals. The main purpose of thesauri is to provide a form of organization with respect to the subject into logical, semantic divisions as well as to index document collections in order to retrieve them. Moreover, basic controlled vocabularies have been developed to reduce ambiguity also by defining terms with scopes notes, while more complex vocabularies provide a set of synonyms for each concept (Tudhope and Binding 2007). The final goal of such a system is to guide final users – through a semantic structure – to someone else's terminology that in specialized domains is required to be extremely specific.

To date, the health data ecosystem comprises a wide array of complex heterogeneous data sources. A wide range of clinical, health care, social and other clinically relevant information are stored in these data sources and in this sense the challenge is to find the right way to access these data in a conscious way. The semantic net on which thesauri are implemented allows to connect concepts to each other in a meaningful way, especially when they refer to heterogeneous data. In this paper we present a source that semantically connects the impact of pollutants on human health.

Pollutants are substances, solids, liquids or gases, mainly produced in high concentrations from human activities, that exert adverse effects on the environment, by polluting the water, the air and the soil (Manisalidis et al. 2020). The negative impact of pollution on human health has been extensively demonstrated. The toxic effects exerted by pollutants are strictly related to their physical and chemical properties. Aerosol compounds have a greater toxicity than gaseous compounds because their tiny size increases their penetration capacity (Colbeck 2009). Many pollutants act as major factors in human diseases, such as particulate matter, nitrogen oxide, carbon monoxide, sulfur dioxide, Volatile Organic Compounds, dioxins and Polycyclic Aromatic Hydrocarbons. The short and long exposure to these pollutants can promote respiratory and cardiovascular diseases, reproductive and central nervous system dysfunctions, cutaneous diseases and cancer, significantly increasing the mortality rate (Manisalidis et al. 2020). Persistent organic pollutants (POPs) include many industrial chemicals with long half-lives, from a few weeks to a dozen years or more, depending on physiological conditions as well as on the environment. The lipophilic properties and the stability of POPs, allow them to persist in soil, air, water and in animal tissue, where they can accumulate and biomagnify.[2] Exposure of human and animals to POPs can lead to a variety of adverse effects, including carcinogenicity and teratogenicity, genotoxicity, reproductive- and endocrine-disrupting effects, immunotoxicity and neurotoxicity (Kailun et al. 2021). Furthermore, heavy metals are included within the group of highly emitted contaminants and their adverse effect on living organisms has been widely studied in recent decades (Houessionon et al. 2021). Mercury is one of the most toxic heavy metals, which can migrate around the globe and magnify through the food chain, ultimately harming human health (Yang et al. 2020).

2.0 The method

In this section the methodological tasks followed towards the construction of a thesaurus for the EO systems domain of study is described. This knowledge organization tool has been aimed at formalizing the terminological arrangement of a specialized lexicon and at providing a semantic resource able to support the indexing operations as well as the retrieval of documentations on this specific reference area of study. The thesaurus creation process is a horizontal activity meant to improve the semantic interoperability and information management procedures for the Earth observation systems domain, and it represents one task included within the two projects, Igosp, part of the ERA-PLANET main program, and E-SHAPE. Igosp focuses on the identification and classification of pollutant hot-spots, while E-shape on global earth sustainability models.

2.1 Corpus compilation

In order to obtain a list of representative technical terms to be used as a starting point for the creation of a semantic network proper to thesauri, the first phase of the activity covered the retrieval of the scientific literature about the domain of study (Trigari 1993). Therefore, the initial procedure started by compiling the source corpus, the collection of documents from which to start to process the textual contents and extract the representative

---

2   Stockholm Convention. The POPs, available at: <http://www. pops. int/TheConven tion/ ThePOPs/ tabid/ 673/ Default.aspx>.

terms meant to be inserted in the thesaurus and be connected with each other by means of the semantic relationships stated by the ISO norms (ISO 25964:2011 and ISO 25964:2013). The documents which have been gathered are included in Pubmed as data sources. This web-based portal uses the The Medical Subject Headings (MeSH) thesaurus to index scientific papers and automatically maps terms to find the ones included within the MeSH controlled vocabulary. In particular, documents have been selected starting from 2000 to 2019 and they refer to original scientific articles (including in vitro and in vivo studies), case reports, systematic reviews and meta-analysis of epidemiological studies.

The approach followed to perform an exhaustive data mining research (Ecker 2010; Bramer 2018) has been based on the establishment of a clear and focused question, the identification of the key concepts addressing the different elements of these latter, and of the elements that should be used to obtain the best results and ordered them by their specificity and importance to determine the best search approach.

In particular, terminologists and domain-experts agreed on the following wildcards criteria (Mishra et al. 2009) to achieve a reliable source corpus for the technical EO domain: (i) identification of the appropriate terms and synonyms as well as combined terms with connector units (AND, OR or NOT); (ii) detection of experts' profiles in the mercury toxicity field to search for their most cited articles; (iii) selection of terms or combined terms to execute the investigation, such as, *environment contamination, occupational human exposure to mercury, mercury and human health, methylmercury OR MeHg AND human health, mercury toxicity, mercury OR MeHg AND cardiovascular diseases, mercury OR MeHg AND neurodegenerative diseases, mercury OR MeHg AND endocrine diseases, mercury OR MeHg AND respiratory diseases, mercury and fetal toxicity, pregnancy AND mercury OR MeHg, molecular mechanism of mercury toxicity, global mercury OR Hg cycle, mercury OR MeHg and Kidney diseases, mercury AND tubular renal toxicity, mercury AND endothelial damage, methylmercury OR MeHg AND brain, methylmercury OR MeHg AND placenta, anthropogenic mercury sources, mercury OR MeHg bioaccumulation, mercury AND urine, mercury tissue accumulation, mercury exposure AND cancer, gold mining AND mercury, dental amalgam mercury, atmospheric mercury concentrations, atmospheric mercury emissions, Minamata disease, Minamata convention.*

A total of 250 studies have been reviewed and they represent the source corpus used to run the next steps related to the semantic analysis and creation of a list of candidate terms useful to build a specialized knowledge-domain thesaurus.

2.2 Text processing

Once having finalized the population of the domain-oriented documen-tation, the following task has concerned the text processing operations to achieve a precise controlled terminological list to analyze. The source corpus has been processed with the aid of a semi-automatic term extractor tool, TextToKnowledge (T2K) (Dell'Orletta et al. 2014), with which the most rep-resentative terms have been identified and filtered out by means of the doc-ument inverse statistical measures. The level of granularity the EO candidate terms bear with respect to this field of knowledge has undergone a valida-tion phase with the experts of the domain who set a specificity treebank for the term base. Indeed, as Trigari (1993) in one or more stages underlines, terms included in thesauri should be selected according to their compli-ance with the controlled language rather than the natural one, meaning that terms need to have morphological features that correspond to standard con-ventions widely used in the documentary procedures, and that they have to be unambiguous dealing with the area of study. The author also points out that term inclusion in thesauri should start by taking into account their *economic value* within the thesaural structure based on exhaustivity and spec-ificity criteria, synonymy preferences, the use of notes which could demark their semantic scope in those cases where the hierarchical and associative structures are not explicitly functional.

The semantic resource developed is the result of the terminological analy-sis starting from the controlled lists of candidate lexical units obtained with the execution of T2K linguistic analysis. In particular, this software is based on the exploitation of Natural Language Processing (NLP) functionalities, statistical text analysis and machine learning methods, and reflects the do-main-oriented information. The output provided by this semantic software is a list of the terms sorted according to their frequency level within the documents and it represents the first step to select the candidate lexical units meant to become part of the higher structured system of a thesaurus (Hud-on 2009). Indeed, the terminologist's activity in identifying the salient infor-mation in the domain-oriented texts is facilitated starting from a controlled list of terms representing the key domain concepts accompanied by their frequency scores retrieved according to their occurrences in the source texts. Other elements to take into account when selecting the candidate terms according to Houdon are linked to the intrinsic scientific value of the terms, their structural compatibility, their contextual dependency degree and co-herence. With the latter criterion the author underlines how the selection of the terms should follow the same logic, meaning that if the terminologists in accordance with the domain-experts start choosing a vulgar form for the representation of the concepts instead of the scientific one, this methodology should be pursued along the entire thesaural construction. In the specific use case presented in this paper, the preferred configuration has been driven towards the systematization of the terminology in the thesaurus framework from a scientific perspective.

2.3 Thesaurus construction

The thesaurus has been chosen with respect to other KOSs (Souza 2012) because of its semantic fixity in defining the associations among domain-representative terms, i.e., managing the unstructured technical information on contaminant substances in an entangled and controlled network of semantic relationships of hierarchy, synonymy and association that can support the understanding of the specialized type of information.

> "[...] thesauri consist of a selection of concepts supplemented with information about their semantic relations (such as generic relations or ('associative relations"). Some words in thesauri are "preferred terms" (descriptors), whereas others are "lead-in terms." The descriptors represent concepts. The difference between "a word" and "a concept" is that different words may have the same meaning and similar words may have different meanings, whereas one concept expresses one meaning." (Hjørland 2007, 367).

Indeed, thesauri represent semantic tools that systematize the information proper to technical fields of knowledge in an interconnection of semantic relationships. The method applied for the construction of the thesaurus has followed the principles contained in the ISO 25964:2011 and ISO 25964:2013 standards where a detailed overview of the thesaurus' construction process starting from the definition of its semantic relationships system is provided. The three main types of relationships refer to hierarchy, equivalence and association connective structures (Broughton 2008). The hierarchical relationship is based on the "degrees or levels of superordination and subordination, where the superordinate concept represents a class or whole, and subordinate concepts refer to its members or parts" (ISO 25964:2011, 58) and it supports the identification of suitable levels of specificity when looking for a more general or more specific term representing a domain-oriented concept. The ISO standard tags for this kind of interconnection are Broader Term (BT) and Narrower Term (NT). An example in the developed EO thesaurus can be the following: BT *contamination*, NT *mercury contamination* NT2 *atmospheric mercury contamination*.

The equivalence relationship stands for the synonymy link to be detected within the knowledge domain, they are tagged as Use (USE) and Used For

(UF) where the first relates to the preferred entry term while the second to the synonyms, e.g. USE *mercury*, UF *Hg*.

Still according to the ISO standard, the associative connection, marked as Related Term (RT) refers to terms that are semantically or conceptually linked to each other but not hierarchically, e.g., *waste incineration* RT *incinerators*.

Following the terminological extraction process the analysis has covered the selection of the main representative terms to be considered as the best candidate entries to include in the thesaurus and to interconnect according to the standards' guidelines. This phase involves a strict form of collaboration with the domain experts responsible for providing a high-level support in the identification of the salient and updated information, in the sense of usability within the specific groups of experts in the domain, to take into account (Shultz 1968).

From a semantic point of view, the specific goals addressed within this activity rely on a formalization of the terminology proper to the EO field of study. The treatment of the specialized languages commonly refers to multiple operations dealing with the semantic disambiguation, with the definition of terminological specificity level and univocity in the way concepts are represented, thus providing a fixed and unambiguous model to be shared with a community of users.

The thesaurus is exploited as a search engine tool to retrieve the documentation needed on the environmental domain focusing on the impacts of mercury and atmospheric pollutants on human health (Aronson 1994). This is how the resource is structured: 979 terms have been included with different levels of depth according to the types of relations set out.

2.4 Semantic interoperability

The perspective of this paper addresses the improvement of semantic interoperability, usually granted by the exploitation of the markup languages or by consulting external vocabularies dealing with the same thematic thread.

For this specific field of study there are several knowledge organization systems available and officially shared by the experts of this domain. For instance, as comparative standards to which to refer to execute the mapping operations, aimed at evaluating the semantic coverage reached in the proposed terminological asset, the existing knowledge organization resources General Multilingual Environmental Thesaurus (GEMET),[3] AGROVOC

3   "GEMET," last accessed September 28, 2021, https://www.eionet.europa.eu/gemet/en/themes/.

Multilingual Thesaurus,[4] Earth and Inspire have been taken into account. Table I shows the best coverage result obtained with GEMET thesaurus (*gemet*) compared to the terminological extraction hereby proposed (*igosp*), and this represents an encouraging result since this latter contains technical information normalized as access points to understand the domain of earth, as the thesaurus developed for this research study yearns to be. The score obtained by mapping the term extraction list with Agrovoc (*agrovoc*) is also encouraging considering that it covers a multidisciplinary range of agricultural sectors: the list from which the EO system thesaurus started to create its internal structure reached a quite low level of the recall measurement, and this should prove the specificity encapsulated with respect to the Agrovoc wide breadth framework.

Corpus term consistency = 93852		
Standards	**n. terms**	**Recall**
Earth	13968	15,8%
Inspire	558	9,8%
Agrovoc	45500	6%
Gemet	5526	20%

Table 1. Coverage with standards

The compliance with the informative texture included in the standards implies a reliable adaptation of the thesaurus to the knowledge domain to be shared, but there are other methods that have been implemented in order to guarantee a deeper level of interoperability. A first factor that impacted the enhancement of the interconnection degree covered the significant collaboration with the experts of this field of study from an environmental perspective as well as medical-scientific one. This interdisciplinary framework facilitated the selection of appropriate terms and semantic structures that could be disseminated throughout several sectors of interest increasing the intra-inter communicability among users with specific skills.

Moreover, the thesaurus has been forged with the objective of creating a resource able to help the indexing of documents, and to be integrated, in this sense, on the GOS⁴M platform (http://www.gos4m.org). In order to implement the thesaurus on this web service, it has been converted into SKOS-RDF language since the flexibility provided by this grammar allows systems

---

4   FAO, "AGROVOC," last accessed September 28, 2021, https://agrovoc.uniroma2.it/agrovoc/agrovoc/en/?clang=fr.

Terms 979		
Terms for each depth level	**Deep level**	**# of terms**
	Deep level 1	673
	Deep level 2	190
	Deep level 3	28
	Deep level 4	8
	Deep level 5	1
Relations between terms 952		
Non-preferred terms 96		

Figure 1. Thesaurus elements

to communicate with each other and facilitate the access to the information (Van Assem et al. 2004).

3.0 Results

In this section some representative examples of the outputs obtained from the previous tasks are presented.

a. Terms and semantic relationships

The thesaurus for EO systems domain is currently constituted by 979 terms and it contains 952 structures of semantic relationships configured through the synergic work of the scientific experts and terminologists, a summary is provided by Figure 1 above.

The next three figures (Figure 2, Figure 3, Figure 4) depict the branching of the semantic connections starting from a preferred term at the top of the lines (*POPs, aqueous mercury, mercury*).

In each tab for selected terms the visualization can be dynamic by expanding the entry terms with their more specific relations, thus elaborating the hierarchical relationship functionality to retrieve knowledge-domain information (McMath 1989), or having a more consistent and flatter alphabetical list where it is possible to see all the relations with the other terms.

93

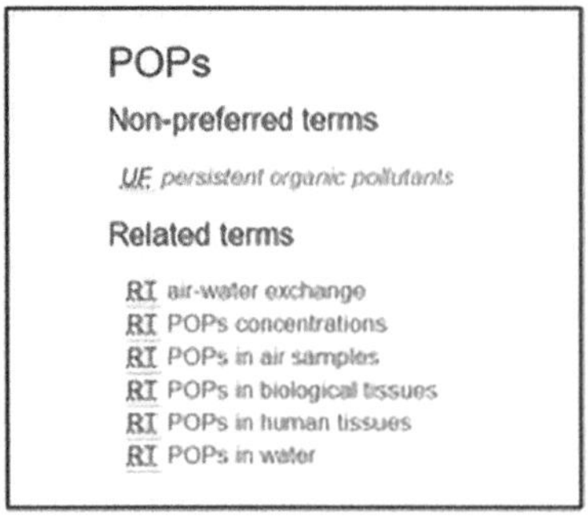

POPs

Non-preferred terms

UF persistent organic pollutants

Related terms

RT air-water exchange
RT POPs concentrations
RT POPs in air samples
RT POPs in biological tissues
RT POPs in human tissues
RT POPs in water

Figure 2. POPs term-set

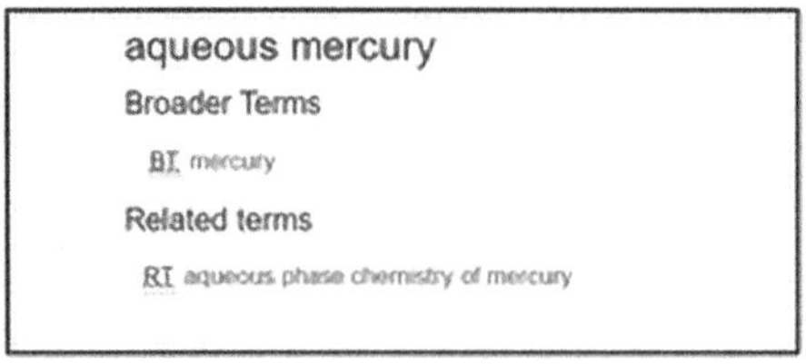

aqueous mercury

Broader Terms

BT mercury

Related terms

RT aqueous phase chemistry of mercury

Figure 3. Aqueous mercury term-set

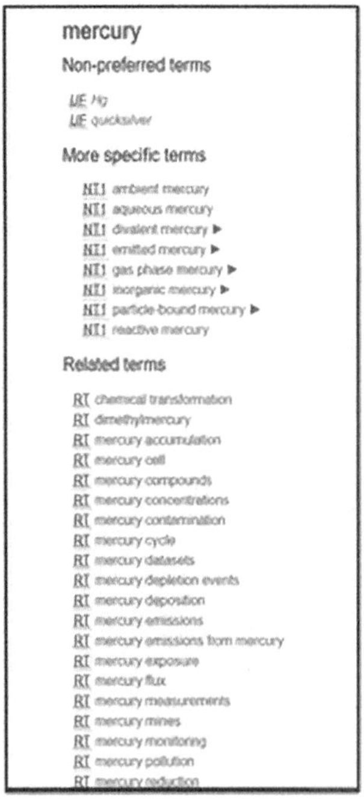

mercury

Non-preferred terms

UF Hg
UF quicksilver

More specific terms

NT1 ambient mercury
NT1 aqueous mercury
NT1 divalent mercury ►
NT1 emitted mercury ►
NT1 gas phase mercury ►
NT1 inorganic mercury ►
NT1 particle-bound mercury ►
NT1 reactive mercury

Related terms

RT chemical transformation
RT dimethylmercury
RT mercury accumulation
RT mercury cell
RT mercury compounds
RT mercury concentrations
RT mercury contamination
RT mercury cycle
RT mercury datasets
RT mercury depletion events
RT mercury deposition
RT mercury emissions
RT mercury emissions from mercury
RT mercury exposure
RT mercury flux
RT mercury measurements
RT mercury mines
RT mercury monitoring
RT mercury pollution
RT mercury reduction

Figure 4. Mercury term-set

4.0 Discussion

According to the final goal of the project, for which the tool has been created,[5] the thesaurus is meant to be integrated on the GOS⁴M web-based platform having as its main goal that of helping users in the indexing and retrieving operations over a large set of documentation about the EO domain. Indeed, by exploiting its inner structure through RDF language, the system can run queries extracting texts where the informative needs are satisfied in a more precise way. The advantages in creating and using a resource such as a thesaurus are intrinsically connected to the way it can be employed to retrieve domain-oriented information by relying on its officially accepted, and consequently reliable, structure of semantic connections. In fact, the searches run over a web-platform with the aid of inferences extracted from the hierarchical, associative and synonymous batteries can help users to directly access a precise group of information (Aitchison 2003). The thesaurus has been built – as mentioned above – with terms coming from scientific literature selected from PubMed that contains studies specifically referring to the impact of mercury on human health conditions, giving in this way an added value to the semantic tool.

Thanks to the support of interdisciplinary teamwork, the integration of information referred to the domain-oriented terms has been achieved. Through the help guaranteed by the experts a list of significant verb clusters has been identified in order to investigate within the source corpus the contexts by which these connections would have offered a detailed mosaic of new data to be imported in the thesaural structure. This activity helped to obtain a richer consistency in the terminology asset by relying on a contextual structure belonging to the texts from which the candidate terms of the thesaurus have been retrieved. In this way, the compliance with the documentation has been maintained and, at the same time, enhanced by exploiting the supplementary information in the same texts through targeted verbal constructions that would have specifically led to precise term associations. In some cases, these patterns have created a new configuration of terms, meaning that terms that were not present in the thesaurus (but they did exist within the output list of the terminological extraction) have been included because of their relationships with the preferred entries, now part of the verbal constructions under examination, e.g., FGDs *UF flue gas desulfurization, hazardous waste RT nonhazardous waste, developing brain RT methylmercury toxicity.*

---

5   The semantic source created is part of the activities of the e-Shape project, Grant Agreement number 820852.

Finally, the main goal of this procedure is to elaborate an enhanced model for the associative relationship proper to thesauri configurations, usually ambiguous and not strictly precise in the connections it forges among the domain-oriented terms.

In addition to this we intend to elaborate an enhanced model for the associative relationship proper to thesauri configurations, usually ambiguous and not strictly precise in the connections it forges among the domain-oriented terms. Some examples of new candidate term connections obtained by using verbal pattern banks to improve the specificity of the associative relation are given in Table 2:

Original term (from corpus)	Semantic relationship created	Term connection
inorganic mercury	RT (is transformed to)	methylmercury
deposited divalent species	RT (transformed to)	gaseous elemental mercury
methylmercury	RT (cause adverse effects in)	developing brain
methylmercury	RT (absorbed through)	gastrointestinal tract
toxic metal	RT (Contaminated)	seafood
methylmercury	RT (excreted in)	breast milk
air-sea exchange	RT (has a large impact on)	GEM concentrations
biological samples	RT (used to assess)	mercury exposure

Table 2. Terms connection examples

5.0 Conclusion

Specialized competences are constituted by specialized terminologies (Cabré 1996), and this is a key aspect to bear in mind when dealing with highly scientific domains of study. Indeed, these specific sectors are characterized by a likewise specific lexicon that implies a series of normalization operations to allow this information to be diffused and create a knowledge representation to be reproduced. In this sense, the paper proposed the creation of a terminological resource, a thesaurus, to be used as a baseline composed by technical terms selected alongside the support and agreement of domain-experts that can represent (i) a structured organization of the EO system knowledge and (ii) a reliable starting point for the information retrieval web-based tasks.

References

Aitchison, Jean, Bawden David, and Alan Gilchrist. 2001. *Thesaurus construction and use: a practical manual*. London: Routledge, 2001.

Aronson, Alan R., Thomas C. Rindflesch, and Allen C. Browne. 1994. "Exploiting a Large Thesaurus for Information Retrieval." *RIAO '94: Intelligent Multimedia Information Retrieval Systems and Management* 1: 197-216.

Auger, Alain, and Barriere Caroline. 2008. "Pattern-based approaches to semantic relation extraction: A state-of-the-art." *Terminology* 14: 1–19.

Bernier-Colborne, Gabriel. 2012. "Defining a gold standard for the evaluation of term extractors". In *Proceedings of the Eighth International Conference on Language Resources and Evaluation*, (LREC 2012): 15–18. https://doi.org/10.1075/term.14.1.

Bramer, Wichor M., Gerdien B. de Jonge, Melissa L. Rethlefsen, and Frans Mast. 2018. "A systematic approach to searching: an efficient and complete method to develop literature searches." *Journal of the Medical Library Association* 106, no. 4. https://doi: 10.5195/jmla.2018.283.

Broughton, Vanda. 2008. *Costruire tesauri*. Editrice bibliografica.

Cabré, Maria Teresa. 1996. "Terminology today". In *Terminology, LSP and Translation: Studies in language engineering in honour of Juan C. Sager*, edited by Harold Somers, 15-34. Benjamins Translation Library. https://doi.org/10.1075/btl.18.

Colbeck, Ian, and Mihalis Lazaridis. 2010. "Aerosols and environmental pollution". *Naturwissenschaften* 97:117-31, https://doi.org/10.1007/s00114-009-0594-x.

Dell'Orletta, Felice, Giulia Venturi, Andrea Cimino, and Simonetta Montemagni. 2014. "T2k^ 2: a system for automatically extracting and organizing knowledge from texts." In *Proceedings of the Ninth International Conference on Language Resources and Evaluation* (ELRA 2014): 2062-70.

Ecker, Erika, and Andrea C. Skelly. 2010. "Conducting a winning literature search." *Evidence-Based Spine-Care Journal* 1, no. 1. https://doi: 10.1055/s-0028-1100887.

FAO. "AGROVOC." Last accessed September 28, 2021. https://agrovoc.uniroma2.it/agrovoc/agrovoc/en/?clang=fr.

"GEMET." Last accessed September 28, 2021. https://www.eionet.europa.eu/gemet/en/about/.

Hjørland, Birger. 2007. "Semantics and Knowledge Organization." *Annual Review of Information Science and Technology* 41, no. 1: 367-405. https://doi.org/10.1002/aris.2007.1440410115.

Houessionon, M. G. Karel, Edgard-Marius D. Ouendo, Catherine Bouland, Sylvia ATakyi, Nonvignon Marius Kedote, Benjamin Fayomi, Julius N Fobil, and Niladri Basu. 2021. "Environmental Heavy Metal Contamination from Electronic Waste (E-Waste) Recycling Activities Worldwide: A Systematic Review from 2005 to 2017". *International Journal Environment Research and Public Health* 18, no. 7: 3517. https://10.1002/aris.2007.1440410115.

Hudon, Michèle. 2009. *Guide pratique pour l'élaboration d'un thésaurus documentaire*. Montréal: Les éditions Asted.

ISO 25964:2011, Information and documentation — Thesauri and interoperability with other vocabularies — Part 1: Thesauri for information retrieval.

ISO 25964:2013, Information and documentation — Thesauri and interoperability with other vocabularies — Part 2: Interoperability with other vocabularies.

Kailun, Sun, Yan Song, Falin He, Mingyang Jin, Jingchun Tang, and Rutao Liu. 2021. "A review of human and animals exposure to polycyclic aromatic hydrocarbons: Health risk and adverse effects, photo-induced toxicity and regulating effect of microplastics". *Science of the Total Environment* 773: 145403. http://doi:10.1016/j.scitotenv.2021.145403.

Manisalidis, Ioannis, Elisavet Stavropoulou, Agathangelos Stavropoulous, and Eugenia Bezirtzoglou. 2020. "Environmental and Health Impacts of Air Pollution: A Review". *Frontiers in Public Health* 8, no. 14. http://doi:10.3389/fpubh.2020.00014.

McMath, Charles F., Robert S. Tamaru, and Roy Rada. 1989. "A graphical thesaurus-based information retrieval system." *International Journal of Man-Machine Studies* 31, no. 2: 121-47.

Mishra, Shruti, Sandep Kumar Satapathy, and Debahuti Mishra. 2009. "Improved search technique using wildcards or truncation." *International Conference on Intelligent Agent & Multi-Agent Systems*, 1-4. https://doi:10.1109/IAMA.2009.5228080.

Shultz, Claire K., Richard H. Orr, and Peter Henderson. 1968. *Evaluation of Indexing by Group Consensus*. Washington: Bureau of Research Office of Education, U.S. Department of Health, Education and Welfare.

Sinha, Pradeep K., Sunder Gaur, Bendale Prashant, Mantri Manisha, and Dande Atreya. 2012. *Electronic health record: standards, coding systems, frameworks, and infrastructures*. New York: Wiley-IEEE Press.

Souza, Renato R., Douglas Tudhope, and Mauricio Barcellos Almeida. 2012. "Towards a taxonomy of KOS: Dimensions for classifying knowledge organization systems". *Knowledge Organization* 39, no. 3: 179-92. http://doi:10.5771/0943-7444-2012-3-179.

Tudhope Douglas, and Ceri Binding. 2009. "Faceted Thesauri." *Anxiomathes* 18, no. 2: 18:211. http://doi:10.1007/s10516-008-9031-6.

Van Assem, Mark, Maarten R. Menken, and Guus Schreiber. 2004. "A method for converting thesauri to RDF/OWL." *The Semantic Web – ISWC 2004: Third International Semantic Web Conference, Hiroshima, Japan, November 7-11* 3298, (Springer 2004): 17-31. http://doi:10.1007/978-3-540-30475-3_3.

Yang, Lixin, Yuanyuan Zhang, Feifei Wang, Zidie Luo, Shaojuan Guo, and Uwe Strähle. 2019. "Toxicity of mercury: Molecular evidence." *Chemosphere* 245: 125586. https://doi.org/10.1016/j.chemosphere.2019.125586.

Analyzing clinical processes and detecting potential correlation between CKD and air pollution[1]

Erika Pasceri
University of Calabria, Italy

Anna Perri
University of Calabria, Italy

Giovanna Aracri
Institute for Informatics and Telematics – CNR, Italy

Sergio Cinnirella
Institute of Atmospheric Pollution Research – CNR, Italy

Abstract

Noncommunicable diseases (NCDs) are the most common causes of morbidity and premature mortality worldwide. The National Chronicity Plan of the Italian Ministry of Health classified Chronic Kidney Disease (CKD) as a chronic disease with a significant level of criticality, highlighting that its management has to be conducted through information systems. To this aim, starting from the clinical data collected within the database of the Department of Nephrology, Dialysis and Transplantation (Annunziata Hospital – Cosenza – Italy), we carried out a propaedeutic analysis of the clinical procedures which are commonly adopted in the care process of CKD defining all procedural flows through process modelling techniques. Furthermore, all clinical data analysed, as it has been reported that environmental pollutants can potentially increase the risk of CKD or accelerate its progression, have been used to define a distribution map over the Calabria Region in order to detect the potential correlation between CKD and air pollution.

1.0 Introduction

Knowledge Organization Systems (KOSs) are conceptual structures, which collect standardized terminology that allow managing, retrieving and sharing data, information and knowledge. They traditionally have been built and maintained throughout the world to support subject indexing of li-

---

1   Authors have equally contributed to this work, however Giovanna Aracri particularly focused on "Introduction", Erika Pasceri on "Method", "Discussion" and "Conclusion"; Sergio Cinnirella on "CKD data", "Pollution Data", "Data Integration"; Anna Perri on "Conclusion".

brary resources (such as books, journals, manuscripts, videos, etc.). However, they can also be used for many other functions (e.g. Information Retrieval (IR), domain knowledge representation and organization, metadata assignment, etc.) (Zeng 2008) and for designing sophisticated platforms able to detect, discover and integrate knowledge (Soergel 2009). The acronym KOSs coined by the Networked Knowledge Organization Systems Working Group (NKOS) gathers a set of controlled vocabularies (terminologies, classification systems, thesauri, semantic networks, ontologies, etc.). Their aim is to systematize and make explicit the semantic structure of both: the general knowledge intended as the range of all known subjects and disciplines; and the domain specific knowledge particularly referred to the set of knowledge acquired through studies or experience by experts in a community of practice. The KOSs optimize the effectiveness of the search systems because they actively contribute to increase the degree of Precision and Recall measures, in order to reduce noise due to non-relevant results and improve their accuracy and pertinence. Thus, the combination of KOSs with IR strategies guarantees the match of the specific request made by the user and the real set of information stored within a collection, a database, a platform, etc. As is known, the IR techniques in order to achieve their objectives need to convert the information request formulated in a natural language into a more formal, precise and unambiguous language. Therefore, this means that both the syntax and the semantics are codified and transmitted according to the latest standard in order to accomplish interoperability. The syntax is essential for implementation and control purposes because it ensures to structure data and information by using reliable, good quality and modeled metadata. On the contrary, the semantics deals with word sense and meaning and therefore with the content of metadata. Syntax and semantics are strictly linked. Their combination is very important since it reduces both lexical and syntactic ambiguity due to the use of synonyms or quasi-synonyms, polysemy and homonyms (ISO 25964-1:2011). Automatically detecting and extracting data and information according to specific parameters and attributes in some domain is crucial. Therefore, the combined use of flexible formats such as Extensible Markup Language (XML) and Unified Model Language (UML) and, KOS especially in some domains, where an extremely high quality of data and information is required, is important for generating reliable sources of data and information. KOSs are useful to support indexing, IR, Knowledge organization and representation and metadata assignment and to accomplish further scopes concerning data and information extraction and integration coming from heterogeneous sources and referred to different but closely related knowledge domains, as we envisage to highlight in this paper. The core of this study regards the definition of the procedural flows about Chronic Kidney Disease (CKD) and the analysis of data collected in order to explore potential correlation between CKD and air pollution.

The huge amount of clinical data produced and stored everyday by health providers needs to be collected and managed in a precise way. This enables data to be accurately used, firstly by the scientific community to constantly learn from each other, especially when the knowledge grows so quickly as in the clinical domain, and subsequently in clinical practice (Cardillo 2016). The relevance of using KOSs in the healthcare domain is widely discussed in the literature. This is strictly related to vocabularies, terminologies or classification and coding systems used to better organize and define clinical concepts and to identify access keys to codified data that can thus be combined, manipulated and shared among healthcare professionals (physicians, data analysts, and all the healthcare operators) involved in the process of care. The KOSs in this way allow to structure and represent complex information fostering their correct interpretation and sharing and to use them in an interoperable way (Cardillo et al. 2016). Nowadays information has been the key to a better organization and new developments. The more information we have, the better we can optimally organize data to deliver the best outcomes (Dash et al. 2019). That is particularly true in the clinical because data and in turn information is used to an improve prediction of a particular disease on the basis of certain parameters. The authors analyzed big data in healthcare, highlighting the management and future prospects, stating that:

> "Healthcare is a multi-dimensional system established with the sole aim for the prevention, diagnosis, and treatment of health-related issues or impairments in human beings. The major components of a healthcare system are the health professionals (physicians or nurses), health facilities (clinics, hospitals for delivering medicines and other diagnosis or treatment technologies), and a financing institution supporting the former two" (Dash et al. 2019, 3).

In each level of the entire care process there is always someone who enters and manages data according to the previous phase, so each professional is responsible for it. The outcome is strictly dependent on the accuracy and completeness of the entire process. The digitization of healthcare has strongly contributed to foster the development of these aspects. On the one hand the transposition of all processes (how data have been managed and transferred) into a digital environment that means to strictly apply rules according to standards adopted (such as HL7), on the other hand data integration that must be compliant to coding or classification systems in use (such as the International Classification of Diseases, Logical Observation Identifiers Names and Codes, Anatomical Therapeutic Chemical classification system, etc.). Concomitantly, the adoption of Electronic Health Record (EHR) systems to store and standardize medical and clinical data – slow at the beginning of 21st century but grown significantly after 2009 (Reisman 2017) – contribute to reduce significantly: the growth of additional or redundant examinations; ambiguities caused by handwriting; time and health waste for chronic patients and for the entire clinical management. Benefits in using EHRs on the clinician side lay on the chance to analyze the entire medical history of patients.

  CKD represents a global public health issue, whose prevalence has gradually increased over the past decade because of emerging risk factors, including some environmental chemicals and particulate matter (PM) that worsen the renal function (Nugent et al. 2011; Tsai et al. 2021). The kidneys excrete waste products from the body, therefore they are susceptible to the adverse effects exerted by the toxins and pollutants circulating in the blood. Epidemiological evidences highlighted that environmental pollutants are important factors in the etiology of CKD and in vitro and in vivo studies elucidated some of the molecular mechanisms by which pollutants, in particular, the heavy metals and PM promote kidney damage (Tsai et al. 2021; Cardelis et al. 2014; Kim et al. 2015). The patients affected by CKD have a higher risk of progression to dialysis and cardiovascular mortality as well as cancer (Raaschou-Nielsen et al. 2017). In addition, long-term exposure to PM has been found responsible for damage to the glomerular barrier and the renal tubule, with consequent alteration of renal function (Lue et al. 2013; Xu et al. 2016). In fact, clinical studies have linked dust levels in the atmosphere with CKD (Bragg-Gresham et al. 2018). Hence, CKD is considered a one of the main health problems for the governments of all countries, whose costs are of great magnitude, as it consumes important percentages of the gross domestic product of the countries (Burgos-Calderón et al. 2021). Therefore, it is crucial to implement national prevention programs to reduce modifiable risk factors, such as exposure to environmental pollutants. Furthermore, the implementation of screening population programs is likewise essential for the early detection of kidney damage, as well as standardized protocols towards a better management of the patients affected by CKD, according to its stage. In this *scenario*, during the medical examination, the clinicians should perform a detailed exposure assessment to establish the potential nephrotoxicants exposure involved in the onset of the kidney disease and its progression to the CKD. More interestingly, the collected information should be encoded within the EHR and not be used only for diagnostic but also for epidemiological purposes.

2.0 The method

The eHealth challenges contribute to the centrality of the inclusive approach to patient care. This awareness is rapidly gaining popularity because it generates a series of benefits concerning the quality of the treatment, the therapeutic and diagnosis choices, the reduction of the waiting time, the simplification of the patient management during his pathway care, especially when he is suffering from chronic pathology as CKD. As previously mentioned, the CKD causes a progressive and complete loss of the kidney function and its progression consists in five stages, each of which requires targeted interventions according to gravity and a multidisciplinary team for handling them in the best possible way. The involvement of several types of health professionals (nephrologists, psychologists and nutritionists) according to their specific competences and experience generates a diversified knowledge that needs to be collected, organized, managed and shared with all subjects taking part in the clinical pathway. From a strictly clinical viewpoint, the Diagnostic – Therapeutic – Assistance Paths (DTAP) listed in national and international guidelines and protocols, already provide a standardized framework for managing some certain patient categories. However, the current organization of the health services especially in the hospital context is rather fragmented. This condition can be due to both: the inadequate availability of clinical specialists and the difficulty of accessing and adequately using the tools already existing. In this sense, it is important to take into consideration, in addition to the clinical dimension, also the technical and the semantic ones. These latter aspects, despite their relevance, are too often neglected and considered as marginal in designing a clinical pathway. Indeed, mixing these layers in reengineering the workflow of a clinical process ensures quality of care throughout the course of the disease. This assumption is valid especially when the complexity of the disease is such as to require several expert's consultation and a close cooperation among them. Guidelines and protocols highlight that the performance improvement of a care process also depends on how information are represented, transmitted and shared. It is an important step towards a greater effectiveness and availability of a treatment. Using IT technologies and devices to support the clinical pathway makes a positive impact from different points of view. From a managerial and organizational perspective, benefits concerning the reduction of waiting times, the way in which information are clearly acquired and transferred, the availability and accessibility of a patient's clinical history; whereas from a purely clinical angle, their use supports the equity in treatment, disease monitoring, etc. In this section, we are going to describe the integrated workflow aimed at providing a clear and comprehensive vision of the CKD. A specific focus has been dedicated to the use of the International Classification of Diseases, 9th revision – Clinical Modification (ICD9-CM) codes adopted to identify diagnosis, services and procedures. The conceptualization of this management workflow allows the collection and the querying of large amounts of heterogeneous data in order to support the development of a modular and scalable platform with the ambitious goal of transforming data into knowledge. It is, therefore, essential to plan the clinical activity by using tools and strategies able to manage, monitor and report all the information produced, whenever a patient accesses a health facility in the form of outpatient service, day-hospital admission, and ordinary hospitalization. Readily available and updated information are synonymous with health services efficiency and quality and it is advantageous for doctors and patients alike. From the point of view of document management, the practices in use at the Department of Nephrology, Dialysis and Transplantation (Annunziata Hospital – Cosenza – Italy) were rather fragmented since they were hybrid. In the face of digital documents enriched by metadata and created by using suitable tools and applications several other documents were in paper format. Because of the lack of homogeneity, it was not possible to guarantee a clear information interpretation to determine statistics or check some regularities of the disease. The attempt was to reengineer the information workflow in use and make it interoperable and compliant to national and international standards by considering step-by-step the actors involved in the document type and format and the terminology and nomenclatures. The combination of these three variables allow establishing: who produces, consults, and edits a document; which document is subject to undergoing revisions and which value set has been modified. Documents produced or received by the Department of Nephrology and included in the information workflow may be medical prescriptions, tickets, clinical reports and discharge letters; each of these can be produced inside or outside the hospital by one or more medical specialists. Therefore, in order to support interoperability and make clinical documents and data searchable, it is necessary to employ international clinical standards such as Health Level 7 (HL7), ICD9-CM, Logical Observation Identifiers Names and Codes (LOINC) etc.

To accomplish the general goal, clinical data of 5017 patients coming from the Department of Nephrology's database have been analyzed. The focus was on patients, whose diagnosis has been codified by means of the 585 ICD9-CM code *Chronic kidney disease*. The following steps have been dedicated to build a database containing clinical and air pollution data with the aim to verify any type of correlation among them (Figure 1).

2.1 CKD data

As stated before CKD data have been collected from the database of Department of Nephrology. This basic information included for each patient the place and date of birth, place of residence and any lifestyle issues. All information has been collected strictly in anonymous form and authorized by

Figure 1. Workflow adopted to integrate domain-specific datasets used to establish the correlation between particulate matter pollution and CKD

the Ethics Committee of the Annunziata Hospital in order to comply with the current legislation on the protection of personal data.

The harmonization process consisted in the preparation of a dataset to proceed to the subsequent geocoding phase. In detail and where necessary, addresses were fixed such as follow for example:

from Via Milano n ° 1 *to* Via Milano, 1
from Via R. Salerno, 8 *to* Via Rosario Salerno, 8
from Via Kennedi *to* Via Kennedy

The geocoding process consisted in the transformation of the address name to the geographical position (latitude and longitude). To this aim the open source Nominatim application[2] that searches OpenStreetMap data by address and generates points coordinates was used.

2.2 Pollution data

A PM dataset was collected from NASA's data center Earth Observing System Data and Information System (https://beta.sedac.ciesin.columbia.edu). The historical PM outdoor trend (1998-2016) is reported as Global Annual $PM_{2.5}$ Grids retrieved from MODIS, MISR and SeaWiFS satellites. It consists of annual concentrations (micrograms per cubic meter) of ground-level fine PM (PM with dimension less than 2.5 micron) at a ground resolution of 0.01 degrees (van Donkelaar et al. 2016; van Donkelaar et al. 2018).

In addition, to consider indoor pollution, wood consumption by different climatic belts was calculated following the methodology proposed by Ozgen et al. (2014) and using the last available socio-economic dataset

---

2   "Nominatim," Github, last accessed October 22, 2021, https://github.com/osm-search/Nominatim.

(http://dati-censimentopopolazione.istat.it/Index.aspx): resident population in built-up areas, residential dwellings, number of households, number of residential buildings and data on the consumption of fuel wood.

2.3 Data Integration

By means of QGIS (https://www.qgis.org) an open source Geographic Information System (GIS), specific values of PM at different locations were obtained. The integrated dataset was finally superimposed to a Digital Terrain Model (DTM) to obtain altitudes at each location. The final dataset was finally prepared as a database for future possible elaborations and an example is reported in Table 1:

Variable	Description	Unit of Measure	Example
id_orig	Dataset Id	#	1
id_paz	Patient Id	#	9877-41
provenienza	City of origin	text	Acquaformosa
eta	Age	years	59
com_nascit	Born place	text	Lungro
com_resid	Residence	text	Acquaformosa
cap_resid	Postal code		87010
lat_deg	Latitude	Decimal degrees	39.722071
lon_deg	Longitude	Decimal degrees	16.090241
cod_reg	National stistic code – Region	#	18
regione	Region	text	Calabria
cod_pro	National stistic code – Province	#	78
provincia	Province	text	Cosenza
cod_com	National stistic code – Municipality	#	1
comune	Municipality	text	Acquaformosa
cod_loc	National statistic code – fraction	#	10001
localita	Fraction	text	Acquaformosa
quota_m	Altitude	meters	781
pop_2000_#	Population in 2000	#	1454
pop_2011_#	Population in 2011	#	1145

Variable	Description	Unit of Measure	Example
abit_resid_#	Number of houses	#	480
fam_resid_#	Number of families	#	480
edifici_#	Total buildings	#	693
edifici_usati_#	Inhabited buildings	#	682
edifici_residenti_#	Residential buildings	#	574
firewood_cons_t	Firewood consumption	tons	24.96
energy_MJ	Energy produced by firewood	Mega Joule	481728
PM_indoor_ug	PM from firewood combustion	µg/m³	246.645
pm25_1998_ug	Average PM, 1998	µg/m³	6
pm25_1999_ug	Average PM, 1999	µg/m³	7.2
pm25_2000_ug	Average PM, 2000	µg/m³	5.6
pm25_2001_ug	Average PM, 2001	µg/m³	6.1
pm25_2002_ug	Average PM, 2002	µg/m³	6.1
pm25_2003_ug	Average PM, 2003	µg/m³	7.6
pm25_2004_ug	Average PM, 2004	µg/m³	5.2
pm25_2005_ug	Average PM, 2005	µg/m³	6
pm25_2006_ug	Average PM, 2006	µg/m³	5.7
pm25_2007_ug	Average PM, 2007	µg/m³	4.6
pm25_2008_ug	Average PM, 2008	µg/m³	4.1
pm25_2009_ug	Average PM, 2009	µg/m³	4.7
pm25_2010_ug	Average PM, 2010	µg/m³	4.1
pm25_2011_ug	Average PM, 2011	µg/m³	4.8
pm25_2012_ug	Average PM, 2012	µg/m³	4.9
pm25_2013_ug	Average PM, 2013	µg/m³	3.9
pm25_2014_ug	Average PM, 2014	µg/m³	4.1
pm25_2015_ug	Average PM, 2015	µg/m³	5.9
pm25_2016_ug	Average PM, 2016	µg/m³	4.9

Table 1. List and example of variables included in the dataset

3.0 Results

The examined dataset consisted of 5017 patients (records) mainly distributed in the Cosenza Province. The analysis using the heatmap methodology

Figure 2. Heatmap of patients' distribution in the Province of Cosenza

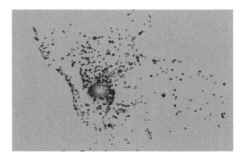

Figure 3. Trend to cluster of CKD patients

(Wilkinson and Friendly 2009) shows some density clusters in the Cosenza area (circular area with a radius of 40 km around the center of Cosenza). The heatmap is a visual representation of data aggregation that encodes the aggregation of information using colors. Figure 2 shows the heatmap of the Province in which some clusters of expected density are highlighted for the urban area and surrounding large towns as Acri and Bisignano.

The analysis of the clustering tendency according to the Hopkins methodology (Hopkins and Gordon 1954) shows a strongly concentrated data structure in the Cosenza area ($h_{mean} = 0.99$) (Figure 3). By carrying out the Optics Clustering (minPoints = 150 and eps = 2450) the data were aggregated to form a single cluster focused on the urban area (Figure 4). By reducing the minPoints value to 70 it is possible to identify minor clusters already highlighted by the heatmap of Figure 5.

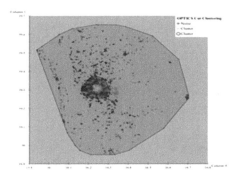

Figure 4. Optics clustering of CKD patients.
Clusters identified with minPoints = 150

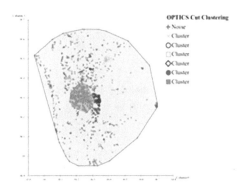

Figure 5. Optics clustering of CKD patients.
Clusters identified with minPoints = 70

A further analysis concerned the distribution of particulate matter. Also in this case a heatmap was produced, which highlighted different levels of pollutants in the various areas (Figure 6). The figure shows as an example the distribution of the concentration for the year 2016.

4.0 Discussion

The effective and complete application of EHRs is an essential prerequisite for the evolution of healthcare systems. It is also known that advancing the notion of clinical data as a public good and a central common resource for advancing knowledge is an evidence for effective care. In addition to this,

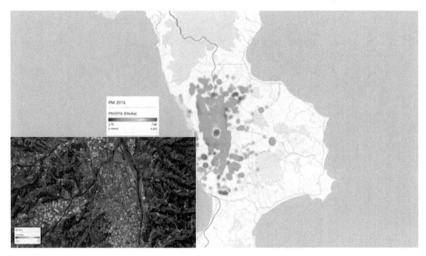

Figure 6. Regional heatmap of the distribution of particulate matter and detail for Cosenza City

obtaining database linkage, data useful for mining and interoperable patient record platforms to foster more rapid learning could better foster consistency and coordination in efforts to better manage knowledge (Institute of Medicine 2007).

In this way clinical decision support systems are considered to be an important vehicle for implementing new evidence and knowledge into daily practice. They generate patient-specific recommendations by matching individual patient characteristics to a knowledge base (Haynes 2010).

Encoded data on diagnoses and procedures are crucial to the success of the health system. Undoubtedly, the correct use of coding systems in the health field supports the management of health information for evidence-based decision-making, primarily for the chronic diseases. Moreover, it allows sharing and comparison of health information between different health facilities, ensuring an appropriate patient care process.

The analysis of the relationship between air pollution and CKD was carried out by considering the distribution of patients and the concentration of particulate matter in the Province of Cosenza. The results obtained by means of statistical analyses highlighted, as expected, the tendency to data aggregation in the urban area of Cosenza but, at the same time, the presence of unexpected clusters such those in surrounding towns. Even in the case of the distribution of particulate matter, clustering in the urban area is evident. The health and environmental data integration can provide unexpected challenges to identify environmental exposures to pollution that may affect human health (Bassil et al. 2015). Data integration and access can enable more punctual decisions through powerful knowledge platforms that bring scientific information prone to the general public and decision-makers.

In addition, the cause and severity of how different environmental pollution and exposures impact human health can be better understood if such integration will be increased, also by fostering data harmonization. There is no perfect methodology to link health and environmental data (IEHIAS, 2021) but the use of GIS is a first step despite its incomplete provision of knowledge (Tim 1995; Sipe and Dale 2003). Much effort is required especially in the collection of harmonized health information that often suffer from fragmentation, clarity and collection methodology (Geneviève et al. 2019). This harmonization and integration can improve data quality, re-usability and interoperability.

5.0 Conclusion

In this study we have investigated the correlation between CKD and air pollution, focusing our attention on $PM_{2.5}$, as experimental and clinical studies demonstrated that the exposure to air pollution, especially PM with a diameter smaller than 10 μm, impairs renal function and increases the risk of incident CKD, leading to end-stage-renal-diseases (Feng et al. 2021). However, several studies demonstrated that other environmental pollutants, including metals and phthalate, exert nephrotoxic effects, potentially increase the risk of CKD or accelerate its progression (Tsai et al. 2021; Wu 2020). In addition, some studies link CKD to intake of metals such as arsenic, cadmium, lead, mercury, copper which have a high nephrotoxic impact (Orr and Bridges 2017; Sun et al. 2019).

The analyses on both environmental pollution and CKD distribution, despite matching, cannot lead to the conclusion that there is a direct relationship. The most significant missing information is the temporal exposure of patients to PM. Moreover, information on metals and phthalate can better address the analysis on CKD-air pollution causal links. Such information not available in our case does not allow a final conclusion on the impact of PM on CKD. Therefore, further studies will be conducted on a larger sample of patients from different areas of the Calabria region, to investigate the correlation of CKD with the environmental pollutants locally detected.

Moreover, an extensive and precise use of coding could better support the decision support systems in analysis focused on heterogeneous data. On the one hand, by using more ICD9-CM specific codes (i.e., 585.3 Chronic kidney disease, Stage III (moderate) to detect specific classes of patients; on the other hand, by using other clinical codings in a complementary manner, for example the LOINC for laboratory reports and the ATC classification system, to detect more information about drug treatments, allergy etc.

References

Bassil, Kate L., Margaret Sanborn, Russ Lopez and Peter M. Orris P. 2015. "Integrating Environmental and Human Health Databases in the Great Lakes Basin: Themes, Challenges and Future Directions." *International Journal of Environmental Research and Public Health* 12, no. 4: 3600–14. https://doi.org/10.3390/ijerph120403600.

Bragg-Gresham, Jennifer, Hal Morgenstern, William McClellan, Sharon Saydah, Meda Pavkov, Desmond Williams, Neil Powe, Delphine Tuot, Raymond Hsu, Rajiv Saran, et al. 2018. "County-level air quality and the prevalence of diagnosed chronic kidney disease in the US Medicare population." *PLoS One* 13, no. 7 e0200612. 31. https://doi.org/10.1371/journal.pone.0200612.

Burgos-Calderón, Rafael, Santos Á. Depine, and Gustavo Aroca-Martínez. 2021. "Population Kidney Health. A New Paradigm for Chronic Kidney Disease Management." *International Journal of Environmental Research and Public Health* 8, no. 13: 6786. https://doi.org/10.3390/ijerph18136786.

Cardillo, Elena, Maria Teresa Chiaravalloti and Erika Pasceri. 2016. "Healthcare Terminology Management and Integration in Italy: Where we are and What we need for Semantic Interoperability." *European Journal for Biomedical Informatics* 22, no. 1.

Dash, Sabyasachi, Sushil Shakyawar, Mohit Sharma, et al. 2019. "Big data in healthcare: management, analysis and future prospects." *J Big Data* 6, no. 54. https://doi.org/10.1186/s40537-019-0217-0.

Feng, Ying-Mei, Lutgarde Thijs, Zhen-Yu Zhang, Esmée M. Bujnens, Wen-Yi Yang, Fang-Fei Wei, Bram G. Janssen, Tim S. Nawrot, and Jan A. Stanssen. 2021. "Glomerular function in relation to fine airborne particulate matter in a representative population sample." *Scientific Reports* 11, no. 14646. https://doi.org/10.1038/s41598-021-94136-1.

Geneviève, Lester D., Andrea Martani, Maria C., Mallet Tenzin Wangmo, and Bernice S. Elger. 2019. "Factors influencing harmonized health data collection, sharing and linkage in Denmark and Switzerland: A systematic review". *PLoS One* 14, no. 12: e0226015. https://doi.org/10.1371/journal.pone.0226015.

Haynes, R. Brian, and Nancy L. Wilczynski. 2010. "Effects of computerized clinical decision support systems on practitioner performance and patient outcomes: Methods of a decision-maker-researcher partnership systematic review." *Implementation science: IS* 5, no. 12. https://doi.org/10.1186/1748-5908-5-12.

Hopkins, Brian, and Gordon John Skellam. 1954. "A new method for determining the type of distribution of plant individuals." *Annals of Botany* 18, no. 2: 213–27.

IEHIAS, 2021. "Integrated Environmental Health Impact Assessment System." Last accessed October 22, 2021. http://www.integrated-assessment.eu.

Institute of Medicine. 2007. *The Learning Healthcare System: Workshop Summary.* Washington, DC: The National Academies Press. https://doi.org/10.17226/11903.

ISO 25964-1:2011, *Information and documentation — Thesauri and interoperability with other vocabularies — Part 1: Thesauri for information retrieval.*

Kim, Ki-Hun, Kabir Ehsanul, and Shamir Kabir. 2015. "A review on the human health impact of airborne particulate matter." *Environment international* 74:136-43.

Maddugari, Santosh Kumar, Vijaay B. Borghate, Sidarth Sabyasachi, and Raghavendra Reddy Karasani. 2019. "A Linear-Generator-Based Wave Power Plant Model Using Reliable Multilevel Inverter." In *IEEE Transactions on Industry Applications*, vol. 55, no. 3: 2964-72. https://doi.org/10.1109/TIA.2019.2900604.

Mehta, Amar J., Antonella Zanobetti, Marie-Abele C. Bind, Itai Kloog, Petros Koutrakis, David Sparrow, Pantel S. Vokonas, and Joel D. Schwartz. 2016. "Long-term exposure to ambient fine particulate matter and renal function in older men: The Veterans Administration Normative Aging Study." *Environmental Health Perspectives* 124, no. 9: 1353-60.

"Nominatim." Github, Last accessed October 22, 2021. https://github.com/osm-search/Nominatim.

Nugent, Rachel, Fathima Sana F., Andrea Feigl, and Dorothy Chyung. 2011. "The burden of chronic kidney disease on developing nations: a 21st century challenge in global health." *Nephron Clinical practice* 118, no. 3: c269-77.

Orr, Sarah E., and Christy C. Bridges. 2017. "Chronic Kidney Disease and Exposure to Nephrotoxic Metals" *International Journal of Molecular Sciences* 18, no. 5: 1039. https://doi.org/10.3390/ijms18051039.

Ozgen, Senem, Stefano Caserini, Silvia Galante, Michele Giugliano, Elisabetta Angelino, Alessandro Marongiu, Francesca Hugony, Gabriele Migliavacca, and Morreale Carmen. 2014. "Emission factors from small scale appliances burning wood and pellets." *Atmospheric Environment* 94: 144-53. https://doi.org/10.1016/j.atmosenv.2014.05.032.

Reisman, Miriam. 2017. "EHRs: the challenge of making electronic data usable and interoperable." *P & T: a peer-reviewed journal for formulary management* 42, no. 9: 572-75.

Raaschou-Nielsen, Ole, Marie Pedersen, Massimo Stafoggia, Gudrun Weinmayr, Zorana J. Andersen, Claudia Galassi, Johan Sommar, Bertil Forsberg, David Olsson, Bente Oftedal, et al. 2017. "Outdoor air pollution and risk for kidney parenchyma cancer in 14 European cohorts". *International journal of cancer* 140, no. 7: 1528-1537. https://doi.org/10.1002/ijc.30587.

Sipe, Neil and Pat Dale. 2003. "Challenges in using geographic information systems (GIS) to understand and control malaria in Indonesia." *Malaria Journal* 2. https://doi.org/10.1186/1475-2875-2-36.

Soergel, Dagobert. 2009. "Knowledge Organization Systems. Overview".

Sun, Yaofei, Quan Zhou, and Jie Zheng. 2019. "Nephrotoxic metals of cadmium, lead, mercury andrsenic and the odds of kidney stones in adults: An exposure-response analysis of NHANES 2007–2016." *Environment International* 132:105115. https://doi.org/10.1016/j.envint.2019.105115.

Tim U.S. 1995. "The Application of GIS in Environmental Health Sciences: Opportunities and Limitations." *Environmental Research*, 71, no. 2: 75-88.

Trigari, Marisa. 1993. *Come costruire un thesaurus*. Franco Cosimo Panini.

Tsai, Hui-Ju, Pei-Yu Wu, Jiun-Chi Huang, and Szu-Chia Chen. 2021. "Environmental Pollution and Chronic Kidney Disease." *International Journal of Medical Science* 18, no. 5:1121-29. https://doi.org/10.7150/ijms.51594.

van Donkelaar, Aaron, and Randall V. Martin. 2016. "Global Estimates of Fine Particulate Matter Using a Combined Geophysical-Statistical Method with Information from Satellites." *Environmental Science & Technology* 50, no. 7: 3762-72. https://doi.org/10.1021/acs.est.5b05833.

van Donkelaar, Aaron, Randall V. Martin, Micheal Brauer, N. Christina Hsu, Ralph A. Kahn, Robert C. Levy, Alexei Lyapustin, Andrew M. Sayer, and David M. Winker. 2018. "Global Annual PM2.5 Grids from MODIS, MISR and SeaWiFS Aerosol Optical Depth (AOD) with GWR, 1998-2016." Palisades, NY: NASA Socioeconomic Data and Applications Center (SEDAC). https://doi.org/10.7927/H4ZK5DQS.

Wilkinson, Leland, and Micheal Friendly. 2009. "The History of the Cluster Heat Map." *The American Statistician* 63, no. 2: 179-84.

Xu, Xin, Guobao Wang, Nan Chen, Tao Lu, Sheng Nie, Gang Xu, Ping Zhang, Yang Luo, Yongping Wang, Xiaobin Wang, et al. 2016. "Long-term exposure to air pollution and increased risk of membranous nephropathy in China." *Journal of the American Society of Nephrology: JASN* 27, no. 12: 3739-46. https://doi.org/10.1681/ASN.2016010093.

Zeng, Marcia Lei. 2008. "Knowledge Organization Systems (KOS)." *Knowledge Organization*, 35, nos 2/3:160 -182. https://doi.org/10.5771/0943-7444-2008-2-3-160.

Wu, Mei-Yi, Wei-Cheng Lo, Chia-Ter Chao, Mai-Szu Wu, and Chih-Kang Chiang. 2020. "Association between air pollutants and development of chronic kidney disease: A systematic review and meta-analysis". *Science of the Total Environment* 706: 35522.

SnowTerm: a terminology database on snow and ice

Paolo Plini
CNR, Institute for Polar Sciences, Italy

Sabina Di Franco
CNR, Institute for Polar Sciences, Italy

Rosamaria Salvatori
CNR, Institute for Polar Sciences, Italy

Abstract

SnowTerm is the result of ongoing work on a structured reference multilingual scientific and technical vocabulary covering the terminology of a specific knowledge domain like the polar and the mountain environment. The terminological system contains around 2.500 terms and it is arranged according to the EARTh thesaurus semantic model. It is foreseen as an updated and expanded version of this system.

1.0 Introduction

The state of the environment and climate change have become crucial and urgent for the scientific, productive and political worlds. It is necessary to combine awareness of the phenomena, correct knowledge of the mechanisms that cause them, and the possibility of effectively mitigating the impacts of human activities.

In all these aspects, correct information is fundamental. Unorganised and uncontrolled data and information can lead to an 'information deluge' with more negative than positive effects.

The organisation of knowledge, starting with a language that is as shared and unambiguous as possible, is one of the central pillars for incisive action to facilitate access to validated and reliable information. Starting with words, concepts and their relationships help to better understand and organise scientific knowledge and reduce misunderstandings. The increasing complexity of scientific and technical subjects forms a network with numerous branches and connections, with frequent overlapping of knowledge domains, which, while being an invaluable asset, complicate the whole system.

Terminological tools (thesauri, glossaries, ontologies) are useful for improving the understanding of domain languages and the organisation of the associated knowledge. Thesauri, in particular, make it possible to share and agree on scientific/technical terms in the target domain and to express them in several languages.

Moreover the use, management and diffusion of information is changing very quickly in the environmental domain, due also to the increased use of the Internet, which has resulted in people having at their disposal a large sphere of information and has subsequently increased the need for multilingualism.

To exploit the interchange of data, it is necessary to overcome problems of interoperability that exist at both the semantic and technological level and by improving our understanding of the semantics of the data. This can be achieved only by using a controlled and shared language.

The polar environment and related polar sciences are no exception to these communication needs.

The polar regions have unique characteristics, such as the phenomenon of polar amplification (Stuecker et al. 2018), which also make them unique in terms of describing their complexity. After research on the internet, several glossaries related to polar and mountain environment were found, written mainly in English. Typically these glossaries – with a few exceptions – are not structured and are presented as flat lists containing one or more definitions i.e. NSIDC Cryosphere Glossary,[1] IPCC "Special report: special report on the ocean and cryosphere in a changing climate Glossary" (Intergovernmental Panel on Climate Change 2019), or the GCW Cryosphere Glossary.[2]

The occurrence of multiple definitions might contribute to increasing the semantic ambiguity, leaving up to the user the decision about the preferred meaning of a term. On the contrary, providing a structure to the lexicon so that each term is placed within a semantic network allows specifying its meaning.

The preliminary results of this work of selection and classification of terms on polar and mountain environment are presented here, as a proposal of controlled and structured language with the goal to develop a prototype of a thesaurus on this specific sector.

The thematic areas, covered at present, deal with snow and ice physics, snow and ice morphology, snow and ice radiometry, remote sensing and GIS in cryosphere environment, sea ice, avalanches, glaciers, disaster management and risk prevention.

---

1   National Snow and Ice Data Center (NSIDC), "Cryosphere Glossary," last accessed October 14, 2021, https://nsidc.org/cryosphere/glossary.

2   World Meteorological Organization (WMO), Global Criosphere Watch (GCW), "Cryosphere Glossary," last accessed October, 14, 2021, https://globalcryospherewatch.org/reference/glossary.php.

2.0 Identification of terminological sources and selection of terms

The first sources used to collect the terminology consist of the "Glossario dei termini usati nei bollettini nivometeorologici",[3] the "Sea Ice Glossary",[4] the "Glossary of Selected Glacier and Related Terminology",[5] the "Sea Ice Nomenclature",[6] the trilingual "Glossary on snow and avalanches",[7] the "Večjezični Slovar – Sneg in plazovi" developed by Pavle Šegula (1995).

The terminology of these sources was analysed with respect to the degree of semantic relevance in the field. Terms too generic or considered as non-pertinent were excluded. Groups of terms that could be collected in specific appendices were also excluded.

At present the database contains 3,700 records; the identification of a certain number of non-descriptors have been performed, the final selection of terms is still ongoing.

3.0 Management of terms

It occurs quite often to find elements belonging to a parent concept which are expressed with terms like "small", "medium", etc. In such cases, we decided to modify the original string adding all the information that will make each term meaningful. For example, the "wind intensity" is declared as "weak", "tempered", "strong", etc. Such terms, if used out of context, are impossible to understand. Having modified these terms in "weak wind", "tempered wind", "strong wind" will allow any user to use the terms in any external application without losing information.

---

3 Associazione Interregionale di coordinamento e documentazione per i problemi inerenti alla neve e alle valanghe (AINEVA) and Friuli-Venezia Giulia Region, "Glossario dei termini usati nei bollettini nivometeorologici," last accessed October 14, 2021, http://www.aineva.it/bolletti/bollet5.html.

4 Scientific Committee on Antarctic Research (SCAR), "Sea Ice Glossary".

5 United States Geological Survey (USGS). 2007. "Glossary of Selected Glacier and Related Terminology," last accesed October 14, 2021, http://vulcan.wr.usgs.gov/Glossary/Glaciers/glacier_terminology.html.

6 "Sea ice nomenclature: English-Finnish-Swedish-Estonian-Russian". *Merenkulkulaitoksen julkaisuja* 5, Finnish Maritime Administration 2002.

7 Swiss Federal Institute for Snow and Avalanche Research, Working Group on Avalanches Warning Services. 2004, "Glossary on snow and avalanches," last accessed October 14, 2021, https://www.avalanches.org/glossary/?lang=it.

4.0 Classification of terms

The classification and relational structure are based on the EARTh (Environmental Applications Reference Thesaurus) semantic model (Mazzocchi and Plini 2005).

The terms are arranged according to a classification scheme that is founded on categories. At the first level, the system is structured into categories defined as "ENTITIES", "ATTRIBUTES", "DYNAMIC ASPECTS" and "DIMENSIONS". The "ENTITIES" describe material and immaterial objects; the "ATTRIBUTES" define the nature of the objects, at least as far as their static aspects are concerned; "DYNAMIC ASPECTS" define the activities, the processes and the conditions in which they are involved; the "DIMENSIONS" identify the spatio-temporal circumstances in which all this occurs.

The system is then organized in a framework of different levels and classification knots, and it comprises hierarchical relations. It continues into further levels as they obtain a greater specificity in order to allow a rational arrangement of objects.

At present around 1,100 terms have been put into the hierarchical structure.

The vertical structure can be used as a semantic reference system, stable and partially independent from the context.

The model envisages the possibility of complementing the faceted structure with a system of themes which by crossing with the vertical structure would form a matrix system.

In a thematic approach, the terms linked to a specific sector are reassembled, while the facet structure tends to scatter them under the more general referral concept.

Moreover, the system of themes, as it was conceived, should be developed by a user according to the specific needs of the applicative context.

One example of thematic setup is provided by the classification into sectors contained in the "Sea Ice Nomenclature" where the terms are clustered according to "ice development", "sky and air indications", "ice arrangement", "terms relating to surface shipping", "terms relating to submarine navigation".

5.0 Software details

All the terms are handled using the TemaTres software (Gonzales Aguilar et al. 2012). It is a tool for visual representation and management of controlled vocabularies. In the frame of an international cooperation some support has been provided to improve the functionalities of the software in particular dealing with the relationships between terms. The web interface allows access to the system through the internet (https://vocabularyserver.com/cnr/ml/snowterm/en/index.php).

6.0 Multilingualism

Multilingualism is not the main interest of our working group. Nevertheless, in order not to waste important resources, the already available translations have been collected. The system now contains 2,700 English terms, the other languages are Italian (2,400), Estonian, Finnish, Russian and Swedish (94), French and German (1,900), Slovenian (1,300) and Spanish (1,800). The enlargement of the number of linguistic equivalents in French and German is mandatory due to the geographical and political position of the alpine area. Other languages will be updated following a direct interest and willingness to cooperate by other institutions.

7.0 Results, their use and future development

The results of this work is the production of a monolingual terminological system organized both in a vertical way -according to a classification system based on categories- and horizontally on the basis of the systems of themes.

SnowTerm could be considered as one of the first attempts to develop a thesaurus on the Polar and Mountain Environment domain.

In order to ensure a better and updated conceptual and terminological coverage, an extension and revision of the system are foreseen. Any other reliable glossary or term list will be considered as potential additional sources.

The semantic structure of the system will also be strengthened. In order to increase the efficiency of the system in information retrieval operations, a set of associative and equivalence relations will be implemented.

The organization of knowledge -through the support of a thesaurus- could bring a strong contribution to the management of the information in the specific domain: by suggesting a language that different institutions could share; ensuring higher semantic transparency to terminology; providing tools for indexing and retrieving the information and to interchange data and suggesting semantic maps usable for the conceptual description of the domain.

References

Associazione Interregionale di coordinamento e documentazione per i problemi inerenti alla neve e alle valanghe (AINEVA) and Friuli-Venezia Giulia Region, "Glossario dei termini usati nei bollettini nivometeorologici." Last accessed October 14, 2021. http://www.aineva.it/bolletti/bollet5.html.

Gonzales Aguilar Audilio, Maria Ramirez-Posada, and Diego Ferreyra. 2012, "TemaTres: software to implement thesauri". *El Profesional de la Informacion* 21: 319-25.

Intergovernmental Panel on Climate Change (IPCC). 2019. "Annex I: Glossary". In IPCC Special Report on the Ocean and Cryosphere in a Changing Climate, edited by N. M. Weyer. https://www.ipcc.ch/srocc/chapter/glossary/.

Mazzocchi, Fulvio, and Paolo Plini. 2005, "Thesaurus classification and relational structure: the EARTh experience". In *7th International conference on Terminology and Knowledge Engineering "Terminology & Content Development"* TKE 2005 16-19 August 2005, Copenhagen, edited by B. Nistrup Madsen, H. Erdman Thomsen, 265-78.

National Snow and Ice Data Center (NSIDC), "Cryosphere Glossary." Last accessed October 14, 2021. https://nsidc.org/cryosphere/glossary.

Scientific Committee on Antarctic Research (SCAR), "Sea Ice Glossary".

Šegula Pavle. 1995. "Večjezični Slovar Sneg in Plazovi. Gorska resevalna sluzba pri Planinski zvezi Slovenije", ISBN: 9616156004.

Stuecker, Malte F., Cecilia M. Bitz, Kyle C. Armour, Cristian Proistosescu, Sarah M. Kang, Shang-Ping Xie, Doyeon Kim, Shayne McGregor, Wenjun Zhang, Sen Zhao, et al. 2018. "Polar amplification dominated by local forcing and feedbacks". *Nature Climate Change* 8, 1076-81, https://biblioproxy.cnr.it:2481/10.1038/s41558-018-0339-y.

Swiss Federal Institute for Snow and Avalanche Research, Working Group on Avalanches Warning Services. 2004, "Glossary on snow and avalanches." Last accessed October 14, 2021. https://www.avalanches.org/glossary/?lang=it.

United States Geological Survey (USGS). 2007. "Glossary of Selected Glacier and Related Terminology." Last accesed October 14, 2021. http://vulcan.wr.usgs.gov/Glossary/Glaciers/glacier_terminology.html.

World Meteorological Organization (WMO), Global Criosphere Watch (GCW), "Cryosphere Glossary." Last accessed October, 14, 2021. https://globalcryospherewatch.org/reference/glossary.php.

Analysis, evaluation and comparison of knowledge extraction tools in the Environmental and Health domain.
A holistic approach

Anna Rovella
Università della Calabria, Italy

Alexander Murzaku
Saint Elizabeth University, USA

Eugenio Cesario
Università della Calabria, Italy

Martin Critelli
Università della Calabria, Italy

Armando Bartucci
Università di Macerata, Italy

Francesca M.C. Messiniti
Università della Calabria, Italy[1]

Abstract

Knowledge extraction in the Environment and Health domains is certainly an important asset for both scientific research and decision support. However, these strategic domains are characterized by a significant heterogeneity of structured and unstructured documents that do not allow a complete transfer of knowledge. Not infrequently, this critical point emerges during the implementation of research projects as an obstacle to the capitalization of information useful for the management of territories or in more recent times to the fight against the pandemic. This paper aims to achieve an analysis of the different forms of knowledge that characterize the scientific production in these specific fields trying to take a holistic approach to text management, tables and graphs through a multidisciplinary logic with the aim of

---

1 Authors contribution statement. Anna Rovella wrote and is responsible for the sections 1.0, 2.0, 3.0, 4.0 and 4.1. Francesca M.C. Messiniti carried out the experiments and is responsible for the statistical data of part 4.1. Alexander Murzaku wrote and is responsible for the section 4.2, Eugenio Cesario wrote and is responsible of section 4.3. Armando Bartucci wrote and performed the experiments of part 4.4 in collaboration with Francesca M.C. Messiniti. Martin Critelli wrote and performed the experiments of part 4.5 in collaboration with of Francesca M.C. Messiniti. All the authors contributed to the concept and design of the study, read and approved the final version of the article.

making the knowledge accessible through a geolocalized representation of sites at risk. A case study on a specific corpus of documents is provided.

1.0 Introduction

Environmental and Earth Observation domains provide a huge volume of heterogeneous research documents. This heterogeneity is also due to the natural intersection with other strategic domains, such as health and agriculture. For this reason, automatic knowledge acquisition and sharing becomes an important asset in this context.

For several years, the research community has worked to build open and shared knowledge bases. Despite these efforts, the need for knowledge extraction tools is still increasing. The relationship between Environmental research and Medical research has accentuated the need for rapid progress on disease-specific knowledge discovery. An example of this correlation is shown when assessing the impact of environmental pollution on the human body.

The large number of sources and the process of knowledge creation make information management a challenging process. In fact, without explicit sharing and effective communication, many data and research results are destined to a very limited use. Instead, an efficient process of knowledge sharing is useful in identifying agents and pathologies more quickly. Furthermore, this process allows a more accurate definition of high impact risk plans with positive effects on the prevention process.

Since the beginning of the pandemic, significant research has taken place aimed at finding new solutions to stop the spread of the contagion. The starting point of the research, which makes it possible, is the observation and collection of environmental data. In the Health and Environmental domain, the selection and definition of knowledge extraction tools, within a holistic vision, are essential for the efforts of researchers and decision-makers for creating and maintaining a non-hostile environment for humankind.

The purpose of this work is to analyse, evaluate, and compare tools for knowledge extraction from scientific literature specifying the described domain. In particular, the evaluation process aims to elaborate quantitative and qualitative data.

An integrated approach requires the identification of critical issues that documents, data and information contained within the Environmental and Health domain. This is a field characterized by structured and un-structured sources, textual documents defined on several levels and which include also objects such as tables and chart images.

The proposed approach aims to overcome some frequent problems in the informationextraction process from the reference literature. These include the extraction of content (metadata, keywords, entities, concepts, objects) but also the possibility of using experimental data often present in tables or images whose information is not immediately understandable and searchable.

To accomplish this goal, we start from the analysis of the performance of some knowledge extraction tools implemented for different and more specific purposes: extraction of metadata, keywords, terms and phrases, tables, charts, and images. The aim is the selection of a class of tools useful for the analysis of all dimensions of the content in the research documents. All the tools exploited in this work use machine learning or deep learning techniques along with different types of analysis and classification algorithms. The comparison and evaluation of all selected tools will be carried out on a specific set of domain documents. Special attention will be paid to the tools that show improved accuracy on the semantic level. The representation of the extracted data is the last step of our work. The purpose of this task is, for example, a possible use of the extracted data in decision support processes. The idea is to represent some extracted data geographical hotspot form and to return the images of tables or the charts in machine-readable form.

The rest of the paper is organized as follows. Section 2.0 presents work related to knowledge extraction from text, table, charts, metadata extraction and knowledge representation. In Section 3.0, we define the methodology of tests and evaluation followed by Section 4.0 in which the results of analysis, evaluation and comparison processes is discussed and an example of representation of geo-referenced data is presented. Finally, conclusions are drawn with notes on limitation and future research efforts are anticipated.

2.0 Related Work

Several techniques for automatic metadata extraction have been studied in the literature and various approaches have led to the implementation of many tools or frameworks. However, in the case of textual documents, structure is more complex. Most of the current tools are built to recognize and classify the basic structure of the input. The information extraction is defined as the identification of entities in the textual content within the document and the relationships of such entities with each other. In this domain, significant results have been achieved through the use of Machine Learning techniques (Liu et al. 2017).

The main limitation of Machine Learning techniques being the absence of data to train and finetune the classification models, we think that metadata extraction tools could provide data to bootstrap this needed training corpus. In addition, we will evaluate the use of *BERT* (Bidirectional Encoder Representations from Transformers) techniques to enrich and better define this knowledge set. This would allow the improvement of the quality of the information extracted through the use of NLP technologies for parsing, tagging, and entity detection. We will apply and evaluate various NLP tools and packages such as Spacy for more precise and finer-grained analysis to research documents.

Table mining can be based on several approaches. They include table detection, functional analysis, structural analysis and semantic analysis. Each of these tasks can be accomplished through different techniques. Various frameworks for information extraction from tables based on multi-layers approaches with high precision scores have been proposed. For data extraction from chart images there are relatively novel Deep Learning approaches (Liu, Klabjan, and Bless 2019).

Moreover, the automatic extraction of geo-referenced data can play a fundamental role in enriching the knowledge model discovery task described above. For example, locations and places referred in the documents can enable the detection of spatial descriptive models, which could be valuable additional information for the Environmental and Health domain under analysis. This can be done by applying some spatial clustering algorithms for the discovery of geographic hotspots, aimed at detecting regions and areas where events of interest occur in with a higher density than other areas.

3.0 The method

The purpose of this work is to analyse, to evaluate, and to compare tools for knowledge extraction from scientific literature specifying the described domain. The results of this research can be subsequently used in various works and domains. One of these, for example, is the implementation of a platform of knowledge analysis and extraction, also in relation to the development of semantic models for the integration of heterogeneous knowledge.

First we define the corpus of documents for knowledge extraction. The selected corpus has already been validated by domain experts, in the European funded *e-shape* project. During that project, the experts more strongly highlighted the need to extract knowledge from images, tables and graphics. The corpus is a subset of scientific articles, extracted from PubMed (PCM database),[2] and concerning the impact of Mercury pollution on human

---

2   US National Library of Medicine, National Institutes of Health, "PubMed Central," last accessed September 28, 2021, https://www.ncbi.nlm.nih.gov/pmc/.

health. It consists of 85 articles on scientific journals in PDF files format. The subject of the articles is Mercury pollution diseases. The small size of the corpus is advantageous as it allows to manually check the results of the automatic extraction. So we were able to obtain a more precise evaluation of the performance of the tools.

We will focus on Machine Learning state-of-the-art solutions, that promise a more scalable solution and more rapid deployment ability.

First we select the tools on the basis of different criteria:

- tools with available open source code;
- tools of which we could verify the installation;
- tools preferably already applied to the Knowledge Base of PubMed.

The selected tools have been used to test their ability to extract knowledge on the chosen corpus. The results of the extraction have been measured and compared.

Some elements played an important role in the choice of tools. For example, for the extraction of metadata we have focused on an adaptive modular approach already tested on PubMed articles (Granitzer et al. 2012). Our purpose, in this case, was not just to check the performance of the tools. Rather, we were interested in understanding where and how to improve the semantic quality of the extracted content or how to adequately represent the information for immediate readability. Just as, in our holistic approach, it was important to test and evaluate the extraction of information from tables and charts and this we know is not a goal of metadata extraction.

The purpose of the linguistic analysis is limited to extracting shallow facts that can be repurposed later in the construction of a knowledge graph. The knowledge elements extracted were named entities, topics, and relations. Combining the extracted geographic named entities with topics provided the input necessary for the identification of geographic hotspots. Relations were detected through syntactic analysis of the text. Noun phrases and similar text chunks were encoded as nodes and verbs and prepositions as edges. Since the English language (all documents were in English) has a fixed order, the task of positioning nodes to the left or the right of the edge reflected the positions in the text itself. When syntactic labels were available, we modified the model to reflect the syntactic roles of the text chunks therefore having subject as left nodes, verbs as edges and objects as right nodes.

In the field of environmental data analysis, the detection of geographic hotspots is becoming a more and more popular task. In this work, we exploit a density-based clustering algorithm to perform a spatial partitioning of the area under investigation, where each cluster represents a dense region of toxicity due to heavy metal exposure. The density-based notion is a common approach for clustering, whose inspiring idea is that objects forming a dense region should be grouped together into one cluster. In our implementation, this step is performed by applying DBSCAN (Density-Based Spatial Clustering of Applications with Noise) (Ester et al. 1996), a popular density-based clustering algorithm that finds clusters starting from the estimated density distribution of the considered data. We have chosen the DBSCAN algorithm because it has the ability to discover clusters with arbitrary shape such as linear, concave, oval, etc. and (in contrast to other clustering algorithms proposed in literature) it does not require the predetermination of the number of clusters to be discovered. Basically, the algorithm finds clusters with respect to the notion of density reachability among points: a point is directly density-reachable from another point if it is not farther away than a given distance (ϵ) (i.e., is part of its neighborhood) and if it is surrounded by sufficiently many points (minPts). In the considered context, a cluster corresponds to a heavy metal toxicity hotspot. Moreover, to capture the dynamic changing of clusters, we could compute the density of each data point by weighting it through a decay factor which gives less importance to historical information and more weight to recent data. Finally, DBSCAN requires the user to specify the radius of the neighborhood (i.e., ϵ) and the minimum number of objects it should have (i.e., *minPoints*), whose values affect size and density of the discovered clusters. Generally, an optimal setting of its parameters is complex to be achieved and requires specific techniques; nevertheless, such a topic is out of the scope of this paper.

Figure 1 shows a schematic representation of our research idea.

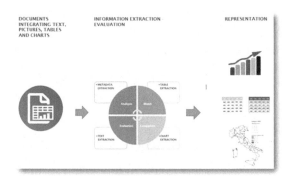

Figure 1. Schematic representation of work

4.0 Analysis, evaluation and representation of extracted knowledge

In this section we carried out the experiments to evaluate the extraction performance of the selected tools on the selected corpus. Each experiment describes processes running, metrics used and results.

4.1 Metadata extraction tool and experimental evaluation.

For metadata extraction we used Cermine (Tkaczyk et al. 2015), a framework created and trained to extract knowledge from PubMed. Cermine is an open source framework for extracting metadata and content from scientific article files in PDF format.

Its modular structure exploits supervised and unsupervised machine learning techniques (Support Vector Machines, K-means clustering and Conditional Random Fields). The System is a prototype developed in java for research purposes and its last update dates back to 2018.[3]

The output produced by Cermine is an xml file in the NLM JATS format.[4] The framework extracts, from documents, mostly Dublin Core metadata (title, author, affiliation, abstract, keywords, journal name, volume, bibliographic references, etc.).

Before describing our experiment, it is useful to recall how Cermine works for metadata extraction operating on the structure of the documents which is analyzed at different levels:

– the characters (dimensions and page coordinates) of the document are read and are identified;
– the different sections of the document are separated by geometric analysis of the pages (page segmentation);
– on the base of page segmentation, character recognition and heuristic structure analysis, the order of reading of the areas of the text is identified;
– then classification process associates metadata and different areas of the text;
– finally, the text is separated from the images, and is classified for the creation of two different output: a file for text and metadata (in NLM JATS format) and a directory for the images (png format files).

---

3 Due to the failure to update the software some framework tools (used for system training) are implemented with deprecated versions of Python 2.6.

4 American National Standard Developed by the National Information Standards Organization (ANSI/NISO). "Z39.96-2012 JATS: Journal Article Tag Suite". NISO, last updated July 26, 2013, https://groups.niso.org/apps/group_public/project/details.php?project_id=93.

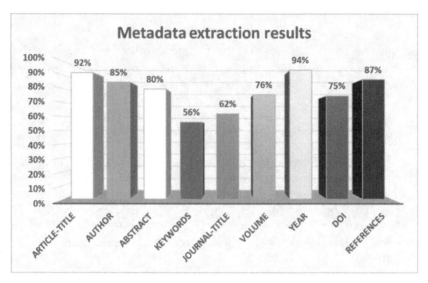

Figure 2. Metadata extraction results

For our experiment of metadata extraction and evaluation we worked according to the following tasks:

– corpus conversion and metadata extraction. Using OCR, we converted the corpus of pdf files into a searchable pdf format. Then we processed it with Cermine 1.13 standalone version. We used the original training set without any personalization;
– metadata extraction evaluation. After metadata extraction we proceeded with the evaluation of the output files by analyzing the quality of the results obtained. For this task we have implemented a specific tool. It is developed in Python with the aim to compare the Cermine output files with the NLM files downloaded from the PubMed Central subset. The tool measures comparison and analysis results by calculating recall and precision scores. Table and diagram form are the output of the tool for showing the metadata quality.

The tool checks the presence of the metadata tags in the file extracted. The values obtained show an extraction rate of more than 50% for all metadata and surprisingly the least extracted metadata are the keywords (56%) while, as can be seen, the year of publication is certainly a data that is always detected (Figure 2).

After this step, the algorithm used in the tool, analyzed the metadata values. This is done by comparing the string extracted by Cermine with the

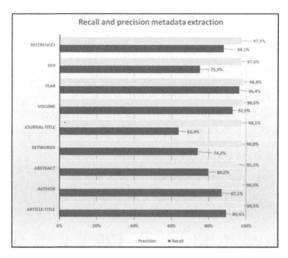

Figure 3. Calculation of recall and precision of meta-data extraction

string stored in the NLM file. For the comparison of the strings and the de-termination of their similarity, the Levenshtein distance (or edit distance)[5] was used. The tool assigns a binary value in case of correct or incorrect ex-traction. The result of the extraction was considered null where the value strings were not complete.[6] These are mainly cases in which the layout mod-els considered by Cermine without customizations are different from those analyzed. In these cases, we obtained a not optimized recognition of the areas of the text. Such situations would require a customization of the layout model that we do not take into consideration in this work. Finally, for each metadata extracted, recall and precision score are measured (see Figure 3). The chart shows that for some extracted metadata such as keywords and journal-title, Cermine obtains a result that is not optimized in quantitative terms (75%) despite the high quality of the information extracted.

From the calculation of the average of the extracted metadata values, we obtained an extraction evaluation with a precision value of 97.9%, recall 83% and error 16.9%.

---

5   "Levenshtein Algorithm," last accessed September 28, 2021, http://www.levenshtein. net/index.html.

6   These are mainly cases in which the layout models considered by Cermine without customizations are different from those analyzed. In these cases, we obtained a not optimized recognition of the areas of the text. Such situations would require a cus-tomization of the layout model that we do not take into consideration in this work.

A consideration comes from the extraction of the body metadata. This is present in a high percentage (96%), with a good quality of the information extracted, even when the article has a different layout from the models of the Cermine training set.

Cermine does not extract data from the images and is not able to recognize, with a good quality, data of the tables.

The evaluation of Cermine's extraction performance is overall positive. However, in order to improve the semantic quality of the extracted metadata, it can be assumed to apply NLP techniques to the analyzed texts. So, we proceeded with the next experiment.

4.2 Knowledge discovery from text

Knowledge discovery from text refers generally to the process of extracting interesting and non-trivial patterns or knowledge from unstructured text documents (Tan, Mui, and Terrace 1999). The research and development of methods, that allow for fast and global analysis of textual data, created conditions for orienting not only decision-making but also the various research disciplines themselves. Given that a large part of scientific research is to understand previous research and to build on it, fast and accurate analysis of published work and knowledge discovery become the focus of attention for many institutions and regulatory organizations.

We are defining text as a general term for sequences of words. Text may be further structured into chapters, paragraphs, or sentences. For our purposes, the text unit that goes through linguistic analysis pipeline is the paragraph marked by the tag "<p>" in the XML output. However, this definition of text includes the concept of "word" which requires a further definition that leads to the concept of token and type. The distinction between a type and its tokens is an ontological one between a general sort of thing and its particular and concrete instances. Thus, 'do', 'does', 'done' and 'doing' are morphologically and graphically marked realizations of the same abstract word type 'do'(Gasparri and Marconi 2021). The process of identifying a token as type is also called lemmatization. Lemmatization usually refers to doing things properly with the use of a vocabulary and morphological analysis of words, normally aiming to remove inflectional endings only and to return the base or dictionary form of a word, which is known as the lemma (Manning, Raghavan, and Schütze 2008, 32). Indeed, in our process of knowledge discovery, the constituents used to identify objects and relations will be lemmas. The last sequence in need for a definition and fundamental for the process of knowledge extraction is the sentence. Sentences according to Quirk et al. (1991) are either simple or multiple. A simple sentence consists of a single independent clause. Subject, verb, complement are constituents of sentences as well as of clauses within sentences. For identification purposes, a sentence is a sequence of words that has boundaries identified by punctuation marks. While question and exclamation points are relatively unambiguous markers of sentence boundary, periods are also present as abbreviation markers such as in etc., Mrs., or Inc. In general, sentence tokenization methods work by first deciding (based on rules or machine learning) whether a period is part of the word or is a sentence-boundary marker.

The text analysis pipeline includes the standard Python (v. 3.9) XML parser (xml.etree) for extracting text paragraphs from the XML files and spaCy (v. 3.1.3) (Honnibal et al. 2020) for tokenization/tagging, parsing/chunking, and named entity recognition (NER).

Text Pipeline

Through the tokenization process, we found that the corpus of 85 documents contains 165596 words/tokens, 16055 sentences, and 21135 unique words/types. Stop words, digits (when possible),[7] one-character long tokens, and punctuation signs are removed from these statistics.

Term frequencies

As expected, since the corpus is composed of articles researching mercury exposure, the list of the most frequent words includes ['mercury':3240, 'exposure':2359, 'study':2082, 'Hg':2059, 'level':1477, 'blood':907, 'high':866, 'concentration':759, 'population':747, 'health':665, 'group':661, 'fish':654, 'biomarker':636, 'child':628, …]. On a per-document basis, TF-IDF (Ramos 2003) is a better distinguisher of relevance as observed in this analyzed document where relevant words are sorted by TF-IDF score: [{doc:0, keywords: {tumour: 80.8, topsoil: 80.5, cancer: 67.5, mortality: 55.8, soil: 39.8, mainland: 34.6, metalloid: 32.3, town: 30.2, spain: 27.2, heavy: 26.3}}]

---

7   The presence of punctuation signs inside a sequence of digits (⸴ and ⸴) is ambiguous on whether it is a decimal point or not depending on the locale.

Named Entity Recognition

The next step in linguistic analysis pipeline is the identification of named entities. Named entity recognition (NER) is the task of finding entities, such as people, locations, and organizations, in text.

The spaCy-recognized named entities are dominated by organizations/institutions (ORG: 6701), numerals (CARDINAL: 6516), dates/periods (DATE: 1940), countries/cities/states (GPE: 1920), and people names (PERSON: 1720). However, the use of a generic named entities recognizer causes a chemical formula (MeHg – methylmercury) to be recognized as a top GPE tag (China: 114, Japan: 62, US: 55, MeHg: 54, USA: 54, Spain: 52). Chaining a specialized NER package such as Chemlistem (Corbett and Boyle 2018) would allow distinguishing of domain specific terms. Also, the NER process exposes the need for coreference identification since in the top twenty GPE list we have US, USA, the United States, and U.S.

Topic detection

Topic detection is a useful mechanism for identifying various concepts embedded in a document, thus, allowing the user to navigate the collection of documents guided by topics. Topics are made up of relevant words, and they provide the user with an overview of the content of the individual documents as well as the document collection as a whole. Since in our sample of articles only 57% have a list of keywords (average 5.4 keywords per article), generating topic related lists of keywords becomes a useful corpus description instrument.

The packages we used are the *gensim* (Rehurek and Sojka 2011) package based on Latent Semantic Analysis (LSA) and the transformer based *BERTopic* (Grootendorst 2020).

Using gensim

In *gensim* every document is represented as a semantic vector. Using unsupervised machine learning algorithms, *gensim* allows for very fast processing and accurate results. The default number of topics is ten and each of them is illustrated by a cluster of ten words and the corresponding scores.

If we looke at the keyword list of the first article that had keywords, we can compare what is generated by *gensim* and what was entered in the publication:

Publication: ['amyotrophic', 'lateral', 'sclerosis', 'ALS', 'motor', 'neuron', 'disease', 'mercury', 'seafood', 'fish', 'consumption', 'dental', 'amalgam', 'filling', 'case-control', 'study', 'online', 'questionnaire', 'international', 'study']

Gensim: ['mercury', 'filling', 'seafood', 'ALS', 'occlusal', 'control', 'respondent', 'dental', 'exposure', 'current', 'proportion', 'online', 'factor', 'respondent', 'exposure', 'silver', 'eat', 'amalgam', 'questionnaire', 'consumption']⁸

The intersection is evident as highlighted by the underlined words.

Using BERTopic

This solution makes use of a sequence of techniques: it starts with the extraction of document embeddings using BERT (Devlin et al. 2018) and then reducing the dimensionality of embeddings to help the clustering process of the reduced embeddings. The output is a set of clusters of semantically similar documents. The final step is the extraction of representative keywords for each document cluster using Maximal Marginal Relevance (Carbonell and Goldstein 1998).

The keyword sets returned by *BERTopic* differ in the form they are organized from *gensim* even though semantically they cover the same meanings. Below is a list of the first 10 keyword groups out of 59.

['als', 'amyotrophic']

['mercury', 'methylmercury', 'methylamino']

['respondent', 'acknowledgment']

['filling', 'precipitate', 'cement']

['seafood', 'seafoods']

['control', 'motor', 'button']

['0111', '15', '1121', '046', '04', '007', '005', '001', '013']

['group', 'people', 'participant', 'human', 'somebody', 'individual', 'community', 'collect', 'committee', 'volunteer']

['dental', 'amalgam', 'tooth', 'mouth', 'bite', 'oral', 'chew']

The results of the location analysis are combined with the two topic extraction techniques allowing for a grouping of topics (such as those above) combined with the corresponding geographic locations such as ['Australia',

---

8   First topic keywords cluster augmented by keywords in the next topic clusters.

'Basel', 'Canada', 'Helsinki', 'Spain', 'Switzerland', 'USA']. This combination allows for a simple answer to the questions what and where.

Some issues with data

The topic modelling output, which is influenced by the relative frequencies of words (TF) as well as specificity of occurrences in the corpus (IDF), includes some peculiar word clusters. By analyzing them we conclude that repeated strings – and this is correct from the algorithmic point of view – are considered as relevant strings. These peculiar strings are generated by OCR errors (which, unfortunately are epxected), and by unexpected languages in the text. These strings affect the TF-IDF calculations of relevance and, therefore, distort keyword/topic detection results.

1. Number of languages included in a corpus

   For our experiment, we chose 85 English language articles; however, analysis shows a different story. Among the 507 unique characters present in the corpus there are:

   i. Latin characters including accented characters more typical of romance languages: ñ, è, î
   ii. Greek characters: μ, β, κ
   iii. Arabic characters: ت, ل, ن
   iv. Cyrillic characters: д, ы, л
   v. Chinese/logographic characters: 考,地,女

   The source of such strings is observed in the bibliography, location/person names, as well as in scientific formulas in the case of Greek.

2. Text extraction from PDF

   Mathematical and other scientific notation text segments generate a large amount of non-word strings as seen below (first the extracted text and second the screenshot of the PDF original text).

   <p>Let Fij denote the factorial burden for each factor (j) at each centroid area location (i). Assume that the observed number of cases Oi in the Ith area is Poisson distributed, with mean $Ei\lambda i$, where Ei is the expected number of cases in that area and the relative risk λi follows a log-linear model, such that:
   logðλiP ¼ α þ ∑4j ¼1β j Fij þ ∑ δk Socik þ ui þ vi</p>

Let F_{ij} denote the factorial burden for each factor (j) at each centroid area location (i). Assume that the observed number of cases O_i in the Ith area is Poisson distributed, with mean $E_i\lambda_i$, where E_i is the expected number of cases in that area and the relative risk λ_i follows a log-linear model, such that:

$$log(\lambda_i) = \alpha + \sum_{j=1}^{4}\beta_j F_{ij} + \sum_k \delta_k Soc_{ik} + u_i + v_i$$

Figure 4. Screenshot of the PDF original text

An open-source software package like Tesseract[9] would allow the separation of scientific notation areas of the text from the rest of the text flow. This would significantly increase the quality of the extracted topics.

Relations

While extraction of topics and recognition of named entities give us a good view of who, what, and where – all of whom can be seen as nodes in a network – a knowledge network would also need a set of connection lines between these nodes. These lines or edges relate well to what in natural languages is expressed through verbs (and some prepositions). Starting with this assumption, we analyze our document corpus using automatic syntactic analysis.

Since the corpus contains documents in the English language, we take advantage of the order type of this language. Once the VERB at the root is identified, all the chunks on the left are considered to enter some relationship described by the verb in the chunks on the right therefore creating NODE-EDGE-NODE triples. Nodes (or chunks) such as *mercury* (758 occurrences), *exposure* (379), *the study* (91), *fish consumption* (89) relate to other nodes via the edges represented by verbs such as *show* (280 occurrences), *measure* (73), *increase* (53), *analyze* (37). For example, the verb *represent* is at the center of these relationships:

---

9   "Tesseract OCR," last accessed September 28, 2021, https://github.com/tesseract-ocr/
tesseract#license.

'these maps' 'each symbol' 'these 75 countries' 'the cross-sectional studies' 'an example' 'nearly 50%' '48.6%' 'the data' 'terms' 'the number' 'individuals'	**represent**	'the average Hg concentration' 'an individual study' 'classes' 'graduated colours' 'the population subgroup' 'the reference' '4 countries' 'order' 'contribution' 'Republic' 'Korea' 'China' 'Japan' 'the United States'

Figure 5. Example of NODE-EDGE-NODE triples

As we are interested in the intersection of *what* and *where* we conclude with some data analysis focused on entities GPE and LOC identifying their adjacent dependent tokens (subject, verb, or object).

what	where
municipalities	Spain
people	USA
people	Australia
University	Montreal
have	Brazil
fell	Islands
came	States
live	Asia
representing	China
recyclers	India
population	USA

Figure 6. Examples of Entities

Notice that MeHg (discussed above) is found in the top occurrences in this corpus (China: 93, Europe: 58, MeHg: 36, Japan: 31, States: 25, Africa: 23, Spain, 20). The top of the *what*-column includes *children* (21 occurrences), *study* (15), *exposure* (13), *countries* (13), *population* (12), and *levels* (12). This approach allows for identifying both where certain issues are faced as well as what issues a certain location faces.

4.3 Discovery of Geographic Hotspots through density-based clustering, experimental Results

To evaluate the performance and the effectiveness of the proposed approach to discover geographic hotspots in a real-world case study, we carried out an extensive experimental analysis by executing different tests in a real scenario, i.e., a set of documents describing mercury toxicity cases occurred in the world.

As described above, geographic hotspots are detected by applying DB-SCAN. As a first consideration before running the tests, in order to detect high quality city hotspots, it is necessary to tune the key parameters of the algorithm so as to improve performance results. DBSCAN takes in input two parameters, ε and *minPts*, which determines the size of the clusters, as they represent the minimum hotspot density required by an area to be part of a cluster. The bigger ε, the larger is the extension of the dense regions detected: this results in the discovery of large regions that actually are no longer dense. The smaller ε, the smaller the cluster sizes, resulting in a high number of dense hotspots detected that could be (because of their small sizes) not significant for the analysis. For what concerns *minPts*, it affects the density of the clusters, that is, the bigger (smaller) *minPts*, the lower (bigger) the average density of the detected clusters. We present here the results achieved by fixing $\varepsilon = 0.1$ and *minPts* = 4, which have been assessed through several experimental tests and best suits our application scenario and the considered dataset.

We performed pilot tests over the geographic data extracted from the documents and following the process described in Sections 4.1 and 4.2. The collected data and the achieved results obtained through our analysis are shown in Figure 7, 8 and 9.

In particular, Figure 7 shows the collected data (right side) and the discovered geographic hotspots (left side) about heavy metal issues discovered in Asia. Each hotspot is represented by a different color. Interestingly, this image shows how heavy metal issue events are clustered on the basis of a density criteria; for example, the algorithm detects several hotspots clearly recognizable through different colors: a large region (in red) in the top-right territory of China, along with several smaller areas (in green, blue and light-blue) on the left (China) and bottom (Japan) sides.

Figure 8 shows the collected data (right side) and the detected geographic hotspots (left side) discovered in Europe. There are clearly recognizable points covering Spain, Italy, United Kingdom, France and Denmark. Also in this case, the algorithm detects several hotspots identified by different colors: a large region (in light-blue) in the bottom-right territory of Spain, along with other areas diffused all over Europe.

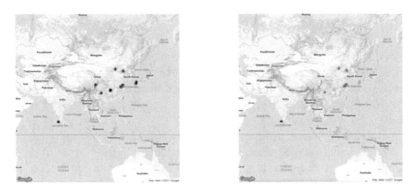

Figure 7: Asia: mercury issues data and detected clusters

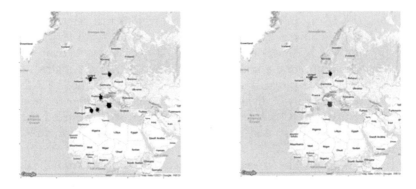

Figure 8. Europe: mercury issues data and detected clusters

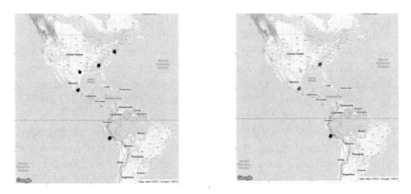

Figure 9. North and South America: mercury issues data and detected clusters

Finally, Figure 9 shows the data and the detected geographic hotspots discovered in North and South America. In particular, several regions in the United States, Mexico and Peru are detected as hotspots of events, representing regions to be considered interesting for further analysis.

4.4 Table Extraction

There are several approaches for table extraction. Each of these approaches can be accomplished through at least two tasks:

- analysis of semi-structured documents based on mark-up language format (e.g., HTML or XML). Tags or coordinates of tables are computed to extract information from tables. However, the scientific articles rarely are already available in mark-up language format;
- pdf files conversion in semi-structured documents format based on HTML or XML languages. PDF is a widely used format in the scientific community for the production of articles. The PDF format does not provide information on the embedded physical layout.

However, it is not easy to convert unstructured documents into semi-structured ones. The main weakness of table extraction by converting PDF files depends on recognizing and understanding tables for automatic tools. This depends on:

- PDF format does not preserve the information related to the document's structure and the structure of tables. This information must be retrieved automatically from the way in which the text content is displayed. Furthermore, most automatic information extraction tools from scientific articles are developed on document layout analysis task based on machine learning algorithms. However, there is no single layout for scientific articles layout. For example, Kise (2014) identifies six kinds of document layout classes. Starting from this classification, Manhattan and Multi-Column Manhattan could be considered the most popular layout used in scientific articles but tables are collocated in different or in more text zones. This determines the dependence of the machine learning model on the single layout used with a consequent negative influence on the extraction results in the case of small layout differences;
- tables may have a different layout defined by the authors or based on the indications of the publishers. The absence of an international standard defining the rules for the creation of tables complicates the recognition and understanding of the model for machine learning-based extraction systems. For example, Luo et al. (2018) observe that the tables in the bio-

medical literature are often presented in a standard form of three-line tables and three-lines of information: *Caption*, *Header field* and *Data field*;
- the cells content can be heterogenous and can contain numbers or text or both. Furthermore, special characters (e.g. mathematical special character as $\pm$) can be detected in an incorrect way.

Analyzing the tools available for table extraction represents an important task to define the state-of-art and propose possible future paths to improve information extraction from scientific articles and better knowledge dissemination.

The next paragraph proposes the analysis, evaluation and comparison tasks of knowledge extraction tools from tables in PDF document. During the experiments we made a comparison between the results obtained by extracting knowledge from tables using a special tool (Tabula) and using CERMINE, a metadata extraction framework. The results obtained from the extraction show the need for specific instruments. But let us describe the experiment.

Analysis and evaluation

The analysis and evaluation activity involved the search for tools for the automatic extraction of information from tables, the technical analysis, the evaluation of the advantages and disadvantages and finally the choice of tools. In the analysis step the main tools useful for this purpose have been identified. Most of them are not free, they do not support all kinds of operative systems, or they are not available because the link is not indicated in the articles. So, we chose Tabula, that has been implemented on a web browser in which the user can upload PDF file containing data table and browse the pages to manually or automatically detect tables by clicking and dragging to draw a box around the table. Tabula will extract data and it will allow to user to select final format (e. g. *.xls or *.csv). Manual detection improves the quality of the extraction but decreases the overall analysis time of the corpus, while the automatic detection decreases time, but the quality of the extraction is lower than manual detection. In this sense, Tabula could be useful to analyze a small corpus of documents (as in this case). For all these reasons, Tabula and CERMINE have been evaluated as the most suitable tools for the phase of extracting knowledge from the tables.

Comparison

The comparison of table mining has been based on the results obtained. We observed that the results can be classified on: "Totally extracted tables",

"Partially extracted tables" and "Not extracted tables". A total of 232 tables have been extracted from the 85 articles of the corpus. Below, we report, an analysis of the results obtained with each software:

- CERMINE

The percentage results on the knowledge extraction from tables with CER-MINE are classified in about 2% of the tables totally extracted, about 28% of the tables partially extracted and finally about 70% of the tables not extracted.

- In 4 cases the tables are completely extracted, and the results are immediately human readable;
- The errors of "Partially extracted tables" class are related to the extraction of a part of table (e.g. extraction of a single column or extraction of only the attributes of the columns) or not all tables in articles are extracted and also in this case they are partially extracted for the same reasons explained before. In this sense, "Partially extracted tables" cannot be read as good results;
- The errors of "Not extracted tables" class are related to the lack of extraction of tables present in the articles.

The reasons of negative results may depend on the strong dependence on the layout of scientific articles. In fact, CERMINE is trained on a Manhattan Layout model but some of the articles are based on the Multi-column Manhattan layout. Furthermore, CERMINE is unable to read tables if shown horizontally in articles and if placed on background texts (e. g. watermark).

The positive results regard the class of "Totally extracted tables". Although they do not contain specific characteristics required by NLM-JATS, all information is presented and could be used in the future for their conversion to NLM-JATS. Furthermore, an important result presented in most of "Table partially extracted" is extraction references. In many cases, CERMINE extracts the information on the references in NLM-JATS format by creating a link with the references section contained in the final XML.

- Tabula

The percentage results on the knowledge extraction from tables with Tabula are also classified in "Tables extracted totally", "Tables extracted partially" and "Tables not extracted" out of the total number of tables in the scientific articles analyzed equal to 232. In particular 87% of the tables were fully

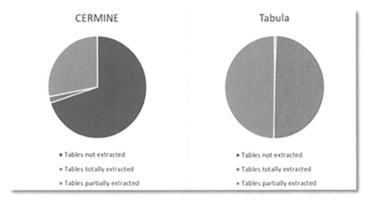

Figure 10. Percentages of "Tables totally extracted", "Tables partially extracted" and "Tables not extracted" using CERMINE and Tabula

extracted, about 12% of the tables were partially extracted and finally about 1% of the tables were not extracted. However, on 87% of the fully extracted tables, about 21% are readable and about 79% are unreadable.

The main limitation of Tabula is the way the information is read. In fact, Tabula reads each line in the tables from left to right and this causes the attributes to overlap between columns if they cover at least two lines of text. This negatively affects the result by determining the high percentage of unreadable tables. However, in this second case the percentage of "Tables extracted totally" tables are higher (caused by the manual detection) than "Tables extracted partially" and "Tables not extracted". It allowed to evaluate the quality of the extracted information. Furthermore, the manual detection makes the tool independent from the layout of the analyzed article but requires time to select each table in the text.

Final Consideration

The comparison considered the percentages obtained for each class of results ("Tables extracted totally", "Tables extracted partially" and "Tables not extracted") and the positive and negative cases of the extraction were used to evaluate the tools. At the end of this comparison, we can establish that Tabula shows a better quality of the information extracted from the corpus of scientific articles. In fact, manual recognition allows you to precisely identify the table in the article and make the tool independent from the layout. However, the use of Tabula is recommended for a limited number of documents. If not, we will spend a lot of time selecting all the tables.

4.5 Data extraction from charts

For charts extraction we analysed ChartOCR (Luo et al. 2021) and ChartReader.[10] ChartOCR is a Deep Learning based framework developed by Luo et al. (2021) for Ubuntu systems and is able to perform data extraction making a Data Table as output. This framework is implemented using CNNs architectures with a Microsoft OCR API to extract text from the image. Since this architecture is very complex, the framework requires a remarkable GPU computing power -in the original experimentation 4 Tesla P100 GPUs were employed. The framework first extracts common information in this case the chart type recognition is performed through the detection of key points. The next phase is the extraction of data range. The data range is calculated in order to read the numerical values inside the graph and, in the final phase, it allows the extraction of data according to the specific type of graph. The last task of the framework is implemented using CNNs architectures with a Microsoft OCR API with the aim to extract text from the image. Since the complex architecture, the framework requires a remarkable GPU computing power.[11] Due to the high GPU required in this work we decided to use another framework, ChartReader, which is characterized by a lighter computer architecture. It is developed by C. Rane and is available on the GitHub page of the author. ChartReader is composed by several modules useful for different purposes:

1. extracting DOI from the PDF articles;
2. recognizing type of chart and axis labels using two different CNNs architectures: VGG-19 & EfficientNetB3;
3. extracting text using AWS API from plots;
4. extracting data.

A last code allows to collect all the extracted information, saved in JSON, inside a single CSV file.

In our experimentation, we use ChartReader as a test on chart images set extracted from the corpus in order to prove their efficiency. The model chosen for testing ChartReader was VGG-19 because its inferior time requirement per inference compared to that required by EfficientNetB3.[12] For the testing phase we used the GPU available on Google Colab. As output we

---

10   Chinmayee Rane, "ChartReader," last accessed September 28, 2021, https://github.com/Cvrane/ChartReader.

11   In the original experimentation Luo et al. (2021) declare having employed 4 Tesla P100 GPUs.

12   For further details visit "Keras," last accessed September 28, 2021, https://keras.io/api/applications/.

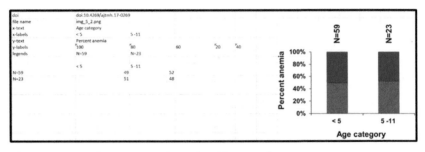

Figure 11: Data & Text extraction using ChartReader

obtained the CSV file which contains all information extracted through the modules described above.

Figure 11 reports a sample of the performed extraction.

5.0 Conclusion and future work

In this paper we proposed a holistic approach to the extraction of knowledge and the representation of information in the environment and health domains. The approach was also exemplified by an experiment conducted on a corpus of 85 scientific papers from PUBMED. The experiment was conducted with a multidisciplinary logic that allowed us, through the application of tools and predictive algorithms, automatic extraction of metadata, text analysis for automatic extraction of content (terms, objects, subjects, entities, relationships), the automatic extraction of data and information from tables and charts, and finally the geolocalized representation of sites at risk. For the experiment we used non customized opensource applications. Although some of the technologies we have used require further optimisation efforts, our approach has shown significant results not achievable on average through one-way approaches. The holistic approach has revealed interesting potential for positive repercussions in the context of research as well as in support of decision-making. The future developments of our research will mainly concern the customization of the tools used and their targeted training also in a logic of integration for the construction of an innovative framework of knowledge extraction.

References

American National Standard Developed by the National Information Standards Organization (ANSI/NISO). "Z39.96-2012 JATS: Journal Article Tag Suite". NISO.

Last updated July 26, 2013. https://groups.niso.org/apps/group_public/project/details.php?project_id=93.

Carbonell, Jaime, and Jade Goldstein. 1998. "The use of MMR, diversity-based reranking for reordering documents and producing summaries." In *SIGIR '98: Proceedings of the 21st annual international ACM SIGIR conference on Research and development in information retrieval August 1998*, 335–6. https://doi.org/10.1145/290941.291025.

Corbett, Peter, and John Boyle. 2018. "Chemlistem: chemical named entity recognition using recurrent neural networks." *Journal of Cheminformatics* 10, no. 59, Springer Nature. https://doi.org/10.1186/s13321-018-0313-8.

Devlin, Jacob, Ming-Wei Chng, Kenton Lee, and Kristina Tautanova. 2019. *BERT: Pre-training of Deep Bidirectional Transformers for Language Understanding*. Cornell University. http://arxiv.org/abs/1810.04805.

Ester, Martin, Hans-Peter Kriegel, Jörg Sander, and Xiaowei Xu. "A Density-Based Algorithm for Discovering Clusters in Large Spatial Databases with Noise." In *Proceedings of the Second International Conference on Knowledge Discovery and Data Mining 2-4 August 1996 Portland, Oregon*, edited by Evangelos Simoudis, Jiawei Han, and Usama Fayyad, 226-31. ISBN 978-1-57735-004-0.

Gasparri, Luca, and Diego Marconi, "Word Meaning", The Stanford Encyclopedia of Philosophy, (Spring 2021 Edition), Edward N. Zalta (ed.), August 9, 2019, https://plato.stanford.edu/archives/spr2021/entries/word-meaning/.

Granitzer, Michael, Maya Hristakeva, Kris Jack, and Robert Knight. 2012. "A comparison of metadata extraction techniques for crowdsourced bibliographic metadata management." In *Proceedings of the 27th Annual ACM Symposium on Applied Computing, (ACM 2012) March 2012 Trento, Italy*, 962-4. https://doi.org/10.1145/2245276.2245462.

Grootendorst, Maarten. 2020. "BERTopic: Leveraging BERT and c-TF-IDF to create easily interpretable topics (Version v0.7.0)." https://doi:10.5281/zenodo.4381785.

Honnibal, Matthew, Ines Montani, Sofie Van Landeghem, and Adriane Boyd. 2020. "Industrial-strength Natural Language Processing in Python." https://10.5281/zenodo.1212303.

"Keras." Last accessed September 28, 2021. https://keras.io/api/applications/.

Kise, Koichi. 2014. "Page Segmentation techniques in document analysis." In *Handbook of Document Image Processing and Recognition*, edited by D. Doermann and K. Tombre, 134-75. London:Springer. https://doi.org/10.1007/978-0-85729-859-1_5.

"Levenshtein Algorithm." Last accessed September 28, 2021. http://www.levenshtein.net/index.html.

Liu, Runtao, Liangcai Gao, Dong An, Zhuoren Jiang, and Zhi Tang. 2017. "Automatic Document Metadata Extraction based on Deep Networks." *Natural Language Processing and Chinese Computing, LNCS* 10619, edited by Huang, Xuanjing, Jing Jiang, Dongyan Zhao,Yansong Feng and Yu Hong. Springer, 305-17. https://doi.org/10.1007/978-3-319-73618-1_26.

Liu, Xiaouyi, Diego Klabjan, and Patrick NBless. 2019. "Data Extraction from Charts via Single Deep Neural Network." arXiv:1906.11906v1.

Luo, Daipeng, Jing Peng, and Yuhua Fu. 2018. "Biotable: A Tool to Extract Semantic Structure of Ta-ble in Biology Literature." In *ICBRA '18: Proceeding of the 2018 5th International Conference on Bioinformatics Research and Applications (ACM 2018) 27-29 December 2018 Hong Kong, Hong Kong*, New York: Association for Computing Machinery, 29-33. https://doi.org/10.1145/3309129.3309139.

Luo, Junyu, Zekun Li, Jinpeng Wang, and Chin-Yew Lin. 2021. "ChartOCR: Data extraction From Charts Images via a Deep Hybrid Framework." In *Proceedings of the IEEE/CVF Winter Con-ference on Applications of Computer Vision 3-8 January 2021 Waikoloa, HI, USA*, 1916-1924. https://doi.org/10.1109/WACV48630.2021.00196.

Manning, D. Christopher, Prabhakar Raghavan, and Hinrich Schütze. 2008. *Introduction to Information Retrieval*. Cambridge University Press.

Quirk, Randolph, Sidney Greebaum, Geoffrey Leech, and Jan Svartvik. 1991. *A Comprehensive Grammar of the English Language*. London: Longman.

Ramos, Juan. 2003. "Using tf-idf to determine word relevance in document queries." In *Proceedings of the first instructional conference on machine learning* 242, no. 1: 29-48.

Rane, Chinmayee, "ChartReader". Last accessed September 28, 2021. https://github.com/Cvrane/ChartReader.

Rehurek, Radim, and Petr Sojka. 2011. "Gensim–python framework for vector space modelling." *NLP Centre, Faculty of Informatics, Masaryk University, Brno, Czech Republic* 3, no. 2.

Tan, Ah-hwee. 1999. "Text mining: The state of the art and the challenges." *In Proceedings of the pakdd 1999, workshop on knowledge discovery from advanced databases* 8, 65-70.

"Tesseract OCR." Last accessed September 28, 2021. https://github.com/tesseract-ocr/tesseract#license.

Tkaczyk, Dominika, Pawel Szostek, Mateusz Fedoryszak, Piotr Jan Dendek, and Lukasz Boli-kowski. 2015. "CERMINE: automatic extraction of structured metadata from scientific literature." *International Journal on Document Analysis and Recognition* 18, no. 4: 317-35. https://doi.org/10.1007/s10032-015-0249-8.

US National Library of Medicine, National Institutes of Health, "PubMed Central." Last accessed September 28, 2021. https://www.ncbi.nlm.nih.gov/pmc/.